Barriers and Fluids of the Eye and Brain

Barriers and Fluids of the Eye and Brain

Edited by

MALCOLM B. SEGAL

Sherrington School of Physiology
UMDS Guy's and St Thomas's Hospitals
London

CRC Press, Inc.
Boca Raton   Ann Arbor   Boston

First published 1992

Published in the USA, its dependencies, and Canada by
CRC Press, Inc.
2000 Corporate Blvd, N.W.
Boca Raton, Fl 33431, U.S.A.

Printed in Great Britain

Library of Congress Cataloging-in-Publication Data

Barriers and fluids of the eye and brain edited by Malcom B. Segal.
  p. cm.
  Includes index.
  ISBN 0–8493–7707–2
  1. Aqueous humor. 2. Cerebrospinal fluid. 3. Blood–brain
barrier. I. Segal, Malcolm B. (Malcolm Beverley)
  [DNLM: 1. Aqueous Humor—secretion. 2. Biological Transport—
physiology. 3. Blood–Brain Barrier—physiology. 4. Carbonic
Anhydrase–pharmacokinetics. 5. Cerebrospinal Fluid—secretion.
6. Prostaglandins—pharmacokinetics. WL 200 B275]
QP476.3.B37 1991
612.8′44—dc20
DNLM/DLC
for Library of Congress   91–20538
                                  CIP

Contents

Contents

The Contributors

D. J. Begley
Biomedical Sciences Division
Kings College London
Strand
London WC2R 2LS
UK

L. Z. Bito
Ophthalmology Research
Columbia University
630 West 168th St
New York
NY 10032
USA

W. Bradley Jr.
Vanderbilt University
Nashville
TN
USA

D. G. Chain
The Weizmann Institute of Science
Rehovat 76100
Israel

J. L. Creasy
Vanderbilt University
Nashville
TN
USA

H. Davson
Sherrington School of Physiology
UMDS Guy's & St Thomas's
   Hospital
London SE1 7EH
UK

F. R. Domer
Dept of Pharmacology
Tulane University School of
   Medicine
New Orleans
LA
USA

J. Fenstermacher
Dept of Neurological Surgery
HSC T12-080
State University of New York
Stony Brook
NY 11794-8122
USA

A. E. James
Vanderbilt University
Nashville
TN
USA

C. Johanson
Dept of Clinical Neurosciences
Program in Neurosurgery
Brown University/Rhode Island
   Hospital
Providence
RI 02903
USA

H. C. Jones
Physiology Section
Division of Biomedical Sciences
King's College London
Campden Hill Rd
London W8 7AH
UK

M. N. Lipovac
Institute of Medical Physiology
Belgrade University
Faculty of Medicine
Belgrade
Yugoslavia

C. H. Lorenz
Vanderbilt University
Nashville
TN
USA

J. A. McKanna
Vanderbilt University
Nashville
TN
USA

K. B. Mačkić
Institute of Biochemistry
School of Medicine
Belgrade
Yugoslavia

R. H. Maren
University of Florida College of
    Medicine
Box J-267
JHMHC
Gainesville
FL 32607
USA

D. M. Maurice
Division of Ophthalmology
Stanford University School of
    Medicine
San Francisco
CA
USA

D. M. Mitrović
Institute of Medical Physiology
Belgrade University
Faculty of Medicine
Belgrade
Yugoslavia

C. L. Partain
Vanderbilt University
Nashville
TN
USA

M. Pollay
Neurosurgical Section
University of Oklahoma Health
    Sciences Center
Oklahoma City
OK
USA

J. E. Preston
Sherrington School of Physiology
UMDS Guy's & St Thomas's
    Hospital
London SE1 7EH
UK

L. M. Rakić
Institute of Biochemistry
School of Medicine
Belgrade
Yugoslavia

N. R. Saunders
Dept of Physiology and
 Pharmacology
University of Southampton
Bassett Crescent East
Southampton
Hants SO9 3TU
UK

M. B. Segal
Sherrington School of Physiology
UMDS Guy's & St Thomas's
 Hospital
London SE1 7EH
UK

E.-P. Strecker
Vanderbilt University
Nashville
TN
USA

J. A. Zadunaisky
Dept of Physiology and Biophysics
New York University Medical
 Center
550 1st Avenue
New York
NY 10016
USA

B. V. Zloković
Division of Neurosurgery
Children's Hospital of Los Angeles
2025 2nd Avenue
Los Angeles
CA 90033
USA

Preface

Over the past 40 years there has been remarkable progress in the understanding of the mechanisms by which the homoeostasis of the fluid microenvironment of the brain and eye is achieved. This period of time spans the research career of Hugh Davson, who, by his own scientific endeavours and those of his many students and colleagues, has played a major role in the understanding of these processes.

This volume reports the latest findings in the field of barriers and fluids of the eye and brain which were presented at a meeting held in London to mark the occasion of Hugh Davson's 80th birthday. Each paper was given by one of Hugh Davson's former students or research colleagues who have worked with him over the years and are now leading experts in the field.

The volume starts with four papers on the vegetative physiology of the eye. The first, by Jose Zadunaisky, reviews the potentials and ion transport mechanisms found at the various interfaces within the eye, and is followed by Lazlo Bito's discussion on the role of prostaglandins in the eye and cerebrospinal fluid (CSF). Both of these topics have been of key interest to Hugh over the years. Maurice Langham and David Maurice were among the first students to train with Hugh Davson and are pursuing distinguished careers in visual physiology. Maurice discusses the problem of the diabetic eye and David gives his own view on the function of the vitreous body.

Tom Maren, who is the 'father' of carbonic anhydrase and its inhibitors, makes a link between the eye and CSF, with a discussion of the role of this important enzyme in both fluids. Michael Pollay, a neurosurgeon who trained with Hugh in his early years and has keen research and clinical interest in CSF, discusses the factors which influence the secretion of this fluid.

Joseph Fenstermacher then enlivens the volume with his new combined method of studying the permeability of local capillary systems within the brain. His chapter is followed by one by Floyd Domer, who takes a pharmacological approach to the effects of lead intoxication on the blood–brain barrier and describes the difficulties of working with this common environmental pollutant.

There then follow two chapters on the permeability of macromolecules across the blood–brain barrier. The first of these chapters, by David Begley and Daniel Chain, reviews the mechanisms regulating peptide levels in the CSF, and is followed by a presentation by Berislav Zloković on the surprising new views on the permeability of the blood–brain barrier to macromolecules.

This chapter is followed by one describing the fascinating morphological studies of Hazel Jones on the development of CSF drainage pathways in both normal rats and those which have spontaneous hydrocephalus.

Norman Saunders then reviews the development of the blood–brain barrier and gives a glimpse of how the comparative physiology of the marsupial can give access to the early stages of the formation of the barrier in free-living neonates, which would be fetal in other mammals.

The final chapter is by Everett Jones (and his co-workers) who, as Professor of Radiology, presents his interesting latest findings on primate CSF drainage pathways. He then completes the major chapters in the volume by giving us a view of the future, with a review of the latest imaging techniques being developed for clinical diagnosis, which will be valuable tools for studying the basic physiology of the fluids of the brain in conscious man, using non-invasive techniques.

At the end of the volume there are a number of short communications from Hugh's young colleagues from Yugoslavia and London who attended the meeting. Hugh has always enjoyed encouraging the young scientist, so no volume dedicated to him would be complete without these abstracts, which hold the promise for the future.

Although I was responsible for the initial idea of this meeting and for much of the local organization, it would not have been possible without the tremendous help and advice freely given by Professor Michael Bradbury at King's College. Mike preceded me in Hugh's laboratory, and it seemed at the time that Mike had already thought of every idea I had for a new project and had performed the experiment during the previous three years!

Mike and I spent a considerable period of time attempting to raise adequate funding, but it is surprising that drug companies, many of which have benefited from Hugh's work, are reluctant to fund basic science, yet will donate large sums for clinical meetings. However, some companies were most helpful and I must thank the following for their generosity: Beechams Research; Imperial Chemical Industries; Smith Kline & French and Pfizer Central Research. We also approached those publishers who have directly benefited from Hugh's writing, but apart from Academic Press, who were most kind, we had no success. In contrast, the Wellcome Trust, without which medical research in the United Kingdom would have foundered, were most generous and we must thank them for making this meeting possible.

We had already invited all of Hugh's closest colleagues who are still busy with research all over the world; they came in spite of our inability to pay their travel costs. The shortfall in funding made us particularly sad, since

Mike and I have been so well looked after in our travels abroad, especially in the USA. We were particularly sorry that this lack of funding meant that Emilio Levin (Argentina) could not be with us.

I began to write some biographical notes on Hugh's career but in the end, after my repeated questions, Hugh decided it would be easier for him to do this himself. This amusing essay follows my introduction.

Hugh has had a considerable influence on research, both through scientific papers and in his books. Those of us who have been privileged to work with him have also enjoyed his friendship and the ability of his keen mind to challenge our hypotheses and fire us with enthusiasm for the next problem.

As you will read, Hugh has often clashed with the more pompous members of the scientific establishment, which for most of us would have spelt the end of our career. Hugh, however, has managed to defeat their machinations, to follow a career dedicated to excellence in research. One of his most endearing characteristics is the help he has always been willing to give to young scientists and to anyone who has come to his laboratory for advice. He has wonderful patience, and will always listen and give his time freely to help others who seek assistance, carefully breaking down the problem in a concise and logical manner. It is a reflection of this ability that the meeting was attended by a large number of young scientists, who came not only for the science, but also to enjoy his lively conversation and company. Hugh has the ability to communicate with all generations and he treats small children as little adults, so that at gatherings of friends and family all the young are found having earnest conversations with Hugh and ignoring the rest of the party.

Communication and language has been his particular forte and he is fluent in French, German and the Romance languages. In recent years, during our co-operation with Yugoslavia, he has learned Serbian for fun, so that when he was elected a member of the Serbian Academy of Sciences, he gave his inaugural address in this, the most difficult of tongues.

I started assisting Hugh with the publication of his many books in 1968 while writing up my PhD. Hugh's first book was *Permeability of Natural Membranes*, published in 1940 with Danielli. This book has been a major influence on permeability studies and is still available today as a reprint. Hugh no longer receives any royalties, as the rights were sold by the publisher to America many years ago. His second book, on the *Physiology of the Eye*, has run through six editions, the latest being published in 1990. Hugh has written a *Textbook of General Physiology*, which ran to four editions, and I well remember helping with the proof reading. Keasley Welch said at the time that Hugh was writing faster than we could read the proof! Hugh's other great interest in the CSF and blood–brain barrier has led to three books on this topic, the latest being *Physiology and Patho-*

physiology of the CSF, and a new volume, a more simple approach to the blood–brain barrier, is in preparation at present. Hugh has also written an *Introduction to Physiology* in five volumes, and edited the multivolume tome *The Eye*. At University College London with Grace Eggleton, he also produced several editions of Starling's *Textbook of Physiology*, which sadly is no longer in print.

I cannot finish without a mention of Hugh's technician, the long-serving Charlie Purvis, who came with him from the Institute of Ophthalmology to University College and built an amazing range of infusion pumps and pressure transducers, without which much of the work could not have been done. The infusion pumps worked at rates down to less than 1 μl/min and were amazingly stable. The equivalent at today's prices costs £3500! The transducers were based on a bronze diaphragm and, being mechanical, used a light lever to record tiny pressures of a few cm of water; with the fine taps he also constructed, they made work in the laboratory a pleasure.

Hugh is still as busy as ever, and as well as writing we are planning some new experiments on an old problem: if successful, we hope to unravel the source of brain extracellular fluid.

We are now discussing plans for the meeting of Hugh's 90th and 100th birthdays, and are fortunate that, although many of us will have 'shuffled off this mortal coil', Hugh's young friends will be there!

Malcolm Segal
London, 1991

Some Autobiographical Notes

Hugh Davson

I was educated at University College School, my father having decided that a change from St. Paul's School, which my elder brothers attended, might be beneficial. I had stayed at my preparatory school a year longer than usual, so that I assumed that I could get along at University College School with no particular effort, as had been the case in my last year at preparatory school. The change from sympathetic to sour and sarcastic teachers, who took pleasure in making fun of pupils, who had no opportunity to answer back, was too sudden and my response was to do no work. My father assumed that I was good for nothing, and so, when I had taken my matriculation, without knowing the result but presuming that I had failed, arranged for me to be taken on by one of his patients, the managing director of a firm operating on the Baltic Exchange. The job was equivalent to an articled clerkship, in the sense that one was unpaid for five years and then became a member of the Exchange, and the going was good for the rest of one's life. The trouble was that there was nothing to learn in those five years that a person of ordinary intelligence could not have picked up in six months' study; after two years, fortunately, I was sacked, presumably for not showing interest! Having passed my matriculation at first taking, a fairly unusual feat then, I was entitled to enter London University, so, as my father agreed to pay the fees, I filled in the necessary form and signed on for chemistry rather than medicine—not that I knew much about chemistry, but I had seen the dog's life a general practitioner led in those days. Thus, I started university life two years after I could have done if I had gone straight from school. In my first year I had to study hard, as I knew little compared with boys who had stayed on at school and failed their Higher School Examinations. I passed these intermediates at a high level and in the second and third years I was up against the cream of scholarship—boys who had passed their Highers and so gained a free place at university. In the final year the results of the examinations taken preliminary to the degree examinations made interesting reading: top were always the triumvirate, usually in alphabetical order, of Brightman, Danielli and Davson. The first two were scholarship types whose brains had survived the ordeal of the scholarship grind. Neither Danielli nor myself could ever beat Brightman, because he had a memory that was completely photographic, and also the intelligence to profit from it. Danielli had the advantage over me of his more thorough training at his secondary school, so that higher mathematics was natural to him, whereas I had had to pick it up as best I could. The results of the BSc examination came out in the predictable order, the triumvirate having firsts.

If times had been prosperous in England at that time, I think the course of my career would have been different: I would have been snapped up by ICI. However, the year of my degree was 1931, when unemployment had reached the three million figure and the Government was retrenching, cutting salaries and even persuading the holders of 5% War Loan to take $3\frac{1}{2}$% voluntarily. Thus, I was 'driven' into an academic research career, in the sense that I had to continue postgraduate work.

I had been fascinated by the science of statistical mechanics, a mathematical approach to physical and chemical problems that was full of possibilities. The Professor of Chemistry at the time was C. K. Ingold, whose lectures on the theory of chemical reactions fascinated me; I felt that if I worked with him, I would be able to realize my intentions of following the statistical mechanical approach. I told him my views and he took me on, but, instead of taking my interest to heart and advising me to study forms of higher mathematics, he pulled out a box file, took out some papers and said: 'This is your project: get on with it.'

The project was the hydrogenation of a conjugate double bond compound which first had to be synthesized; a previous research student had obviously failed and I was to take it on. The synthesis of the compound I was supposed to hydrogenate was very difficult and nobody thought of giving me any help beyond telling me that I should polish my bench more carefully; thus, by the end of this first year of research, I was told that I could not be put forward for a research grant. With no prospect of further financial help from my father, who was on the point of retiring, this created quite a problem: I had married secretly and my wife, Marjorie, was showing signs of reproduction. Having been interested, along with Danielli, in the problem of permeability of cell membranes, having listened to a series of lectures by J. C. Drummond, then Professor of Biochemistry, in which he had described some interesting effects of calcium on the permeability of the estuarine worm *Gunda ulvae*, and having had informal chats with him on several occasions, I went to him, explained my situation and asked him whether he would apply on my behalf for the research grant. I proposed to work on permeability, a subject on which Drummond, a vitamin chemist, had had no experience. As a testimonial to a man I revered and loved more than anyone else in the world, I can say that he took me on and gave me a free hand. I obtained the necessary grant, reduced by government economy from £150 to £120, and I can remember the raptures with which the first quarterly instalment, of £30, was received. A bank account was immediately opened and a cheque for five shillings was sent proudly to a brother, who had probably lent that amount.

I was then left to my own devices. Having dipped into the literature, I had decided that the permeability of red cells to potassium and sodium was interesting, so the problem of chemically determining these ions in organic matter occupied me for a very considerable time.

In spite of considerable difficulties in the chemical analysis of these ions I produced a paper which was published in the *Biochemical Journal* (1); it had been reviewed by Rudolf Hober, the leading German general physiologist, a recent refugee from Hitlerism. Looking back on it, I realize the appalling mistakes I had made in my experimental approach, although, of course, the actual results were impeccable. The mistakes were, I am sure, due to a lack of supervision, but then there were no qualified general physiologists about.

However, to get on with my story, my grant was drawing to a close and I was faced with the prospect of a starving wife and baby if I did not get another grant. Professor Drummond called me into his office one day and said that he had a job for me with Glaxo, at, I think, £300 a year. The research was naturally to be on matters of industrial importance but I was to have some chance of doing research on my own; he said, in all kindness, that I owed it to my wife to take the job. Looking back, I am amazed that I did not do so; £300 a year was six pounds a week, compared with the two pounds ten shillings that we were scraping along on. However, I remember I thanked him and said that my wife would not forgive herself if I abandoned a pure research career, although I am sure I never mentioned the matter to her at all! Interestingly, as a parenthesis, the man who did take the job was Magnus Pyke, who was doing a PhD on vitamins and was obviously more suited to it than myself.

The next development was that a successful ophthalmologist, Stewart Duke-Elder, who prided himself on his contributions to the physiology and biochemistry of the eye, having operated on Prime Minister Ramsay Macdonald for glaucoma, was given a knighthood and then needed someone to continue his researches. I was the someone and was given a grant of £250 a year, which transformed my life from one of complete penury to one in which I could afford to buy a few ounces of tobacco a week; beer, I recollect, was still out of the question.

The immediate problem was whether the cause of primary chronic glaucoma was a swelling of the vitreous body, a theory proposed by Duke-Elder. I naturally assumed that he was right and did my best to prove it, but it became more and more obvious that this was an incorrect hypothesis. I finally managed to explode the idea by obtaining an eye in absolute glaucoma and taking the vitreous out. My argument was that if it was causing a pressure of perhaps 50 mmHg in the eye, it must have a swelling pressure to correspond; in fact its swelling pressure, which I was able to measure by a special device, was less than a few millimetres of water. With my aid, therefore, Duke-Elder was able to show that he had been wrong, in a paper entitled 'The swelling pressure of normal and glaucomatous vitreous bodies' (2). This is an interesting paper: I submitted it to the *Biochemical Journal* but it was rejected on grounds that showed the referee did not understand the problem. By objecting to this, I came into conflict for the first time with the Establishment, this time in biochemistry. The editor was C. R. Harington, and I can remember my Professor, Drummond, calling me in to say that I must make my peace with him because I had said to one of his underlings, whom I had met in a pub, that I thought the criticisms of my paper were stupid! However, being young and independent, peace was not made and I paid for this indiscretion subsequently.

While working on the eye, I also continued my studies on the permeability of the erythrocyte. Danielli had, meanwhile, been in the USA on a Commonwealth Fellowship and had carried out work on the plasma membrane, following up an idea we had had together at University College when trying to explain the effects of calcium on the worm *Gunda ulvae*; we thought that it could be the effect of the divalent ion on the surface film of lipid that would bind the lipid molecules closer together. Danielli wrote the paper that was to have so much influence on thought so far as the cell membrane was concerned (3). In this context, I must say that Jim Danielli was precocious to an uncommon degree; usually the precocious schoolboy gets to a university but after that fades out. Danielli not only got to a university but also obtained a good first and then became a first-class research worker; there was no feeling his way, so that in all my association with him, which lasted from, say, 1933 to 1937, when he went to Cambridge, he was the 'superior spirit'. Together we wrote a number of papers on red cell permeability and laid the foundation for the paucimolar theory of the membrane.

In 1935 I asked Professor Drummond whether I could get a fellowship to spend a year in America; he thought it was a good idea and approached the Rockefeller Foundation, who awarded me a fellowship in 1936. I suppose I had made a small reputation for myself in the field of permeability, and the leading general physiologist, M. H. Jacobs, took me on to do a year with him. Although I did not get much help from him, I did learn that you must repeat your results thoroughly before committing yourself to print. I was actually disappointed by him, because I was interested in the effects of narcotics on permeability; Danielli and I had thought that these should decrease permeability by virtue of their action on the surface film membrane—a concept that dominated our thinking at the time. Jacobs had found that they had, if anything, the opposite effect, but he was working with ox red cells. By chance I chose rabbit red cells and found a profound decrease in permeability to glycerol by ethyl alcohol; however, I was discouraged from publishing these controversial results. I had in fact discovered an important feature of permeability, since it turned out that the facilitated transport type of permeability, of which Danielli and I were just beginning to become aware, was liable to effects of narcotics, whereas the plain, non-facilitated type of transport was not. The non-publication of these data actually retarded progress in the understanding of transport processes.

During my stay in the USA I had applied for a Beit Fellowship, a coveted award that enabled the holder to carry out research for three years with prospects of a senior fellowship for another three years. On returning to London I agreed to continue work on the eye but my main interest was now the cell membrane. The Fellowship began in 1937 and by 1938 Professor Drummond thought that I should find a teaching job, as ultimately I must take one if I was to become a professor, the only possibility of advancement in those days. I successfully applied for the position of Associate Professor at Dalhousie University, Canada. Prior to my departure for Canada, I had Peter Quilliam to assist me; he was a good practical physiologist and acquired the technique of perfusing the isolated cat's head with the Starling heart–lung preparation. Using this perfusion system, I was able to modify the concentration of potassium in the blood and measure the blood–aqueous barrier; this, I suppose, was the first quantitative study in which I developed the two-compartment kinetics that forms the basis of most subsequent work on the ocular and cerebrospinal fluids (4).

I had spent the two previous summers at Cold Spring Harbor, the Director of which, Eric Ponder, was an authority on the red blood cell. He encouraged me to do research with him (5). I went to Canada from New York by boat and then by rail to Halifax; on the train I learnt that Britain was at war.

My chief at Dalhousie was Professor Weld, who was mainly interested in teaching, but that created an ideal situation for me, as all the money available for research was at my disposal. I got Professor Weld to help in a study of the effects of aphakia on the composition of the aqueous humour (6). Meanwhile I learnt a great deal of physiology, as there were only the two of us to do all the teaching of medical students. All this time I was doing my best research work; my day usually consisted of arriving at about 8.30 a.m., giving a lecture from 9 until 10, and then plunging straight into the laboratory. I was not much worried during the period of the 'phoney war', but when the Germans looked like winning, I wrote to my Professor (Lovatt-Evans) in London, asking him to help me to get back to do wartime research; in 1942 I and my family embarked on the *Sarpedon*, a former P & O liner, in convoy for Liverpool. We arrived in Liverpool without any serious incident and began life in England.

I was sent to work at the Chemical Defence Establishment at Porton Down in Wiltshire. Here I found that there was a new gas, nitrogen mustard, that attacked the eyes, so I devised various experimental set-ups with rabbits to measure the effects on the aqueous humour. To this end I brought Peter Quilliam down from London and together we did some interesting work on the effects of nitrogen mustard which has a permanent place in the literature of the aqueous humour (7). After a year or so of Porton, as the threat of gas attacks receded, I felt that I was doing no useful service and I eventually moved to London with the Army Operational Research Group.

AORG was created by the war and used everybody's talent to the best. Because of my experience on the eye, I rather ambitiously took on visual problems and was involved at once with infra-red illumination. You had an infra-red searchlight which illuminated the scene only to those who had the necessary infra-red receivers. It was excellent fun: I learnt to drive heavy army lorries, Sherman tanks, armoured cars, and even bulldozers, all with the aid of infra-red illumination. I also managed to lose a searchlight, and, since I signed for it, I am always slightly worried that one day the MoD will send me a bill!

At the end of the war I was faced with a dilemma: I had been given leave of absence from my job as Professor in Canada, but the university had appointed a temporary occupant of the position and wanted to know whether I intended to return. If I had had any sense, I would have gone back for a short period, because I feel sure now that I would have been offered a number of academic posts. As it was, the thought of Canada was too much for me, so I gave in my resignation and was on my own.

Over the past few years I had been in contact with Duke-Elder and we had discussed plans for an Institute of Ophthalmology in which there would be a department of research, which I would head. We persuaded the Medical Research Council to give us a grant to start some research on the eye at University College. With Duke-Elder I built up a team of some four or five young people who subsequently obtained their PhD degrees on projects initiated by myself. The group consisted of David Maurice, Maurie Langham, A. M. Woodin, E. J. Ross and N. Ambache (8).

When the building for the Institute of Ophthalmology had been converted for its purposes, I moved my team in and work proceeded with tolerable harmony for perhaps a year. However, before long there was a clash of personality between Duke-Elder and me, so I went to the MRC. We all have crises in our lives and this was the most appalling crisis in mine: once again I was flying in the face of the Establishment. Fortunately, Professor Gaddum of University College Pharmacology Department was on the Council, and when the matter was brought up, he was able to state, in no uncertain terms, that I was the guiding light in the whole research unit and that it was scandalous that I should be allowed to leave. However, prejudice in favour of the Establishment prevailed, so instead of my being put in absolute control of the unit (which, I must say, I had not asked for—merely to be released from the Institute), I was allowed to set up myself with one technician, my instrument maker, Mr Purvis, who said he would leave the Institute if he were not allowed to work with me, and a secretary. Thus, I began research life again at a time when I and one or two associates should have been engaged in either developing my theories of permeability or developing theories as to the nature of the relations between blood and the tissues of the eye—namely, aqueous humour, cornea and lens (9,10). These areas have been studied by my younger colleagues with no subsequent help from me, because of my withdrawal, and, of course, no competition, since I felt myself bound to abstain from research in those fields in which I had given these younger colleagues my advice.

It was because of this that I turned to the study of the cerebrospinal fluid. I began studies of the blood–CSF barrier in, I suppose, about 1953, having first completed a study on the blood–aqueous in collaboration with my personal technician, Parnel Matchett, who moved with me, along with Mr Purvis from the Institute of Ophthalmology, in, I think, 1951. After publishing this study with Matchett (11), I studied the blood–aqueous and blood–CSF barriers simultaneously in the rabbit, using partly chemical and partly isotopic techniques, ^{24}Na having become available on a weekly basis. While doing this initial study, which lasted some two years, I decided that I must get to know the literature of the CSF and blood–brain barrier and, as the easiest way to do this, wrote a book. This was *Physiology of the Ocular and Cerebrospinal Fluids* and it came out in 1956, at about the same time as the 1955 paper on CSF and eye fluids (12). (Previously I had published *Physiology of the Eye* in 1949 and *Textbook of General Physiology* in 1951.)

In 1953 I was invited to spend a few months in Rio de Janeiro with Professor Chagas, who had built up a biophysics laboratory in which the electric organ was the main subject of study; I remember Richard Keynes had preceded me. To pass the time I measured the concentration of K^+ in intra- and extracellular fluids and showed that they were consistent with a resting potential of some 50 mV (13).

About this time, Cecil Luck spent a year with me at University College. He had been in Africa and had arrived in England looking for a job; he was in the Department of Physiology, working with young Michael de Burgh Daly, and G. L. Brown, I think, suggested that I take him on. At any rate, I asked him to join me on a comparative study of aqueous humour and CSF in mammals; we were analysing their fluids for chloride and bicarbonate (14). He handled animals well, including monkeys, and we did some very good work. As a result of this, when the chair at McKerrerie in Uganda became vacant, he was able to convince the selectors that he could do some good work on comparative physiology of mammals and obtained the chair.

The first American to spend a year with me was Gene Spaziani, from Los Angeles. He had been trained in endocrinology but his chief thought that he should learn something about permeability. He performed experiments on both the eye and the blood–brain barrier (BBB) (15)—in particular, the work with brain slices that was the first 'brick' thrown at the anatomists, led by my colleague J. Z. Young at University College, who were claiming that the BBB was a fiction.

Later, in 1961, I spent a summer at Wood's Hole doing a little research on the eye fluids of fish; Cynthia Grant and I published a small note in the *Biological Bulletin*. While there, I received a telephone call from Keasley Welch, Professor of Neurosurgery at Denver, who had seen a notice in *Science* to the effect that I was in the country; I spent a few days in Colorado with him, giving seminars in sufficient number to defray the costs of myself and my wife. This formed a basis of a lasting friendship between us, a friendship shared by many of my younger colleagues, so when Keasley moved to Boston Children's Hospital, this lovely city became, and still is, the first stop for many of us on our various tours.

After Spaziani, I think the next American was Chuck Kleeman, also rather more of an endocrinologist than a general physiologist. He made an excellent study on the kinetics of the barrier to urea (16). My next visitor was a young colleague of Keasley Welch's, Mike Pollay. He developed the ventriculocisternal perfusion set-up in the rabbit (17) and subsequently was given the first chair in neurosurgery at Albuquerque. When Pollay left me, the MRC, who had previously refused stubbornly to provide any help in the way of qualified research workers, allowed me to take on Michael Bradbury, who had read my book on the fluids, and asked the new professor at Oxford, G. L. Brown, who had gone there from University College, to arrange this.

In 1963 I was approached, cap in hand, by a Louisville ophthalmologist, Dwight Townes, who wanted to found an Institute of Ophthalmology in Louisville and asked whether I would be the director of research and visiting distinguished professor. I agreed to this, as I could now leave my London laboratory in charge of Bradbury. Meanwhile, Laszlo Bito had written, asking whether he could do a post-doctoral couple of years with me. I told him that I was going to Louisville and that I presumed he would not want to go there. However, he said he would and he arrived a few weeks before I did. I soon discovered that all that was wanted of me was my signature at the bottom of application forms for grants. Thanks to Bito's moral support, I refrained from throwing in the sponge and we spent a profitable year working on the eye and CSF (18); it was there that we did the first 'microdialysis' experiment in an attempt to measure the true extracellular concentration of potassium in the brain, my idea being that it would be equal to that in the CSF. Our membrane sac was large, however, and had to be implanted in the forebrain of a dog and allowed to stay there for some weeks before we could venture in to get the fluid. Within the last few years the technique has been improved out of all recognition, and we are actually setting up the system at St. Thomas's.

While at Louisville I took part in a symposium in Buenos Aires. Here I met Zadunaisky, whom I persuaded to take on my job at Louisville after I left in 1964; I also met again Emanuel Levin, who had worked for a few months with me and had established the technique of ventriculocisternal perfusion in the cat.

After my return to London, Bito came to do a second year and, together with Bradbury, we spent a very exciting period working mainly on the active transport of iodide and bromide across the blood–brain barriers (19). When Bito left, Bradbury went to work in Los Angeles with Kleeman, where he stayed some three years.

In 1967 Bill Oldendorf spent a year with me; his interest had been confined to X-ray studies of the CSF system in humans, so working on rabbits was new to him (20). When he went back to Los Angeles, he devised the now famous BUI technique; I remember his writing to me outlining it and asking whether I approved, and Bradbury and I both encouraged him to go on. In 1971 he sent me a reprint of his first paper on the technique and wrote in the cover: 'See what a year in Davson's laboratory does for you.'

When he left, I was approached by Malcolm Segal, who was taking his finals in physiology, and he spent some three years doing his PhD with me. We did some very good work together, made possible by the unfailing skill of my instrument maker, Mr Purvis (21). We were the first to measure the resistance to drainage of the CSF, a technique that was followed up by the Americans (22).

In about 1970, Dr Joan Abbot wrote to me from the USA, telling me that she had got a 'rehabilitation grant' permitting her and her husband to return to England, having regretted, I presume, falling down the 'brain drain'. At the same time, the MRC suggested that a member of the Carshalton Unit, Dr Ian Glen, might benefit by a year's work in my laboratory. Both Joan Abbott and Ian Glen spent a year with me, after which Joan obtained a position at King's College and Ian Glen one in Edinburgh (23). During this period, Jack Stulc from Czechoslovakia also joined the laboratory, so a most enjoyable and productive time was had by all.

When Segal left to go to St. Thomas's, where Bradbury had taken a job, Keasley Welch decided to spend a year with me; he had taught himself a lot of mathematics and so was able to analyse the blood–CSF, blood–nerve and blood–brain barriers in a manner that I, with my more limited knowledge of mathematics, would not have attempted (24).

In about 1972 Dr Floyd Domer, a pharmacologist at Tulane University, New Orleans, came to do a year's sabbatical in the laboratory (25). Floyd was unusual in that he could drink as much beer as an Englishman!

In South America David Yudelivich was having trouble with the change of government in Chile and I was able to persuade the MRC to support him for a year in the laboratory before he obtained the chair in physiology at Queen Elizabeth College.

I continued working at University College with my technician, Gillian Hollingworth (26), until I was due to retire at about age 66. Meanwhile the Fogarty Foundation had invited me to take a scholarship for a year after retiring, so I went to work at the NIH for a year. During this year I had all the money I wanted to organize a symposium on the Ocular and Cerebro-spinal Fluids, and took pleasure in inviting all my colleagues, including two technicians, to come to it with all their expenses paid. It gave me great pleasure to see them all, especially Mr Evans, who had been chief technician at University College since I began research there in about 1933. He had seen one after another of us going to the USA but never dreamed of being invited there himself, and I think he enjoyed himself very much and looked the most distinguished of all those present!

After my year at Bethesda, during which I spent as much time as possible visiting my former colleagues in different parts of the USA, I returned to London and obtained a three-year grant from the MRC. I had intended to work at University College but changed my mind and moved to King's College in Mike Bradbury's laboratory. There my first visitor was Mike Michaelson from Cincinnati, to study dipeptides (27); next came Mike Carey, a New Orleans neurosurgeon, who studied the effects of insulin hypoglycaemia on the barriers (28); and finally, Joe Fenstermacher, who spent a year studying amino acid transport out of CSF (29).

At the expiry of my grant, I was shocked at the impudence of the MRC, who refused me another one; however, the Wellcome Trust came to my aid and I continued for several more years, roping in to my work David Begley. We employed Danny Chain, the son of the penicillin Chain, as our technician (30); he has now taken a PhD in Israel and is working in the USA. While at King's I had made the acquaintance of Berislav Zloković, who had been doing a stint with Yudelivich at Queen Elizabeth College. Berislav collaborated with me and Malcolm Segal (31) and has now taken up a job with Gordon McComb in Los Angeles. Gordon spent the year 1974–1975 in my laboratory; an excellent neurosurgeon, he was able to cannulate the sagittal sinus and torcula of the rabbit (32). I think it was while Gordon was working with me that Everette James came over and did some work on drainage of CSF. He was fundamentally a radiologist, who had kindly invited me to Vanderbilt when I was at Bethesda.

Of recent years I have been busy with writing a further edition of *Physiology and Pathophysiology of CSF* with Keasley Welch and Malcolm Segal and the fifth edition of *Physiology of the Eye*. Malcolm Segal arranged for me to be a visiting Professor at UMDS St. Thomas's Hospital, and I have an office in the Sherrington School of Physiology. Here I have enjoyed the many visits of our young Yugoslav colleagues, students of Berislav Zloković, who visit London to work in Segal's laboratory supported by the British Council and the kindness of the Wellcome Trust. I had an amusing time disproving, with Jane Preston, Segal's PhD student, my dear friend Tom Maren's finding that aluminium completely inhibited CSF secretion; we found that the observed inhibition was in fact a pH colour artefact and that CSF secretion is only slightly reduced by this ion (33). My current line of research with Sarah Williams, another of Segal's students, involves laying the 'spectre of Stern and Gautier's hypothesis', which will appear in press in the near future.

REFERENCES

1. Davson, H. (1934). Studies on the permeability of the erythrocytes. *Biochem. J.*, **28**, 676–683
2. Duke-Elder, S., Davson, H. and Benham, G. H. (1936). The swelling pressure of normal and glaucomatous vitreous bodies. *Br. J. Ophthalmol.*, **20**, 320–527
3. Danielli, J. F. and Davson, H. (1935). A contribution to the theory of thin films. *J. Cell Comp. Physiol.*, **5**, 495–508
4. Davson, H. and Quilliam, J. P. (1940). The permeability of the blood–aqueous humour barrier to K^+, Na^+ and Cl^- in the surviving eye. *J. Physiol.*, **98**, 141–154
5. Davson, H. and Ponder, E. (1938). The permeability of 'ghosts' to cations. *Biochem. J.*, **32**, 736–762
6. Davson, H. and Weld, C. B. (1941). Studies on aqueous humour. *Am. J. Physiol.*, **134**, 1–7
7. Davson, H. and Quilliam, J. P. (1947). The effects of nitrogen mustard on the permeability of the blood–aqueous humour barrier to Evans Blue. *Br. J. Ophthalmol.*, **31**, 717–721
8. Davson, H., Duke-Elder, W. S., Maurice, D. M., Ross, E. J. and Woodin, A. M. (1949). The penetration of some electrolytes and non-electrolytes into the aqueous humour and vitreous body of the cat. *J. Physiol.*, **108**, 203–217
9. Langham, M. and Davson, H. (1949). Studies on the lens. *Biochem. J.*, **44**, 467–470
10. Davson, H. (1949). The aqueous humour and the blood–aqueous barrier. *Ophthalmic Literature*, **3**, 254–268
11. Davson, H. and Matchett, P. A. (1951). The control of intraocular pressure in the rabbit. *J. Physiol.*, **113**, 387–397
12. Davson, H. (1955). A comparative study of the aqueous humour and cerebrospinal fluid in the rabbit. *J. Physiol.*, **129**, 111–133
13. Davson, H. and Lage, H. V. (1933). The extracellular space and internal K concentration of the electric organ of *Electrophorus electricus*. *Acad. Brazil. Ciencas*, **25**, 303–307
14. Davson, H. and Luck, C. P. (1956). A comparative study of the total CO_2 in ocular fluids, CSF and plasma of some mammalian species. *J. Physiol.*, **132**, 454–464
15. Davson, H. and Spaziani, E. (1959). The blood–brain barrier and the extracellular space and the brain. *J. Physiol.*, **149**, 135–143
16. Kleeman, C. R., Davson, H. and Levin, E. (1962). Urea transport in the central nervous system. *Am. J. Physiol.*, **203**, 739–747
17. Pollay, M. and Davson, H. (1963). The passage of certain substances out of the CSF. *Brain*, **86**, 137–150
18. Bito, L. Z., Davson, H., Levin, E., Murray, M. and Snider, N. (1965). The relationship between the concentrations of amino acids in ocular fluid and plasma of dogs. *Exp. Eye Res.*, **4**, 374–380
19. Bito, L. Z., Bradley, M. W. B. and Davson, H. (1966). Factors affecting the distribution of iodide and bromide in the CNS. *J. Physiol.*, **185**, 323–354

20. Oldendorf, W. M. and Davson, H. (1967). Brain extracellular space and the sink action of CSF. *Arch. Neurol.*, **17**, 196–205
21. Davson, H. and Segal, M. B. (1970). The effects of some inhibitors and accelerators of sodium transport on the turnover of ^{24}Na in the CSF. *J. Physiol.*, **209**, 131–153
22. Davson, H., Hollingworth, G. and Segal, M. B. (1970). The mechanism of drainage of the CSF. *Brain*, **93**, 665–678
23. Abbot, N. J., Davson, H., Glen, I. and Grant, N. (1971). Chloride transport and potential across the blood–CSF barrier. *Brain Res.*, **29**, 185–193
24. Davson, H. and Welch, K. (1971). The permeation of several materials into the fluids of the rabbit brain. *J. Physiol.*, **218**, 337–351
25. Domer, F. R., Davson, H. and Hollingworth, G. R. (1973). Subarachnoid versus ventricular perfusion in the rabbit brain. *Brain Res.*, **38**, 81–94
26. Hollingworth, J. G. and Davson, H. (1973). Transport of sulphate in the rabbit's brain. *J. Neurobiol.*, **4**, 389–396
27. Begley, D. J., Davson, H. and Michaelson, M. A. (1980). Clearance of the dipeptide glycyl-l-glycine from rabbit CSF. *J. Physiol.*, **307**, 83P
28. Carey, M. E., Davson, H. and Bradbury, M. W. B. (1981). Effect of severe hypoglycaemia upon CSF formation. *J. Neurosurg.*, **54**, 370–379
29. Davson, H., Hollingworth, J. G., Carey, M. B. and Fenstermacher, J. D. (1982). Ventriculo-cisternal perfusion of 12 amino acids in the rabbit. *J. Neurobiol.*, **13**, 293–318
30. Davson, H., Begley, D. J., Chain, D. G., Briggs, F. D. and Shepland, M. T. (1986). The steady state distribution of cycloleucine and AIB between plasma and CSF. *Exp. Neurol.*, **91**, 163–173
31. Zlovokić, B. V., Segal, M. B., Begley, D. J., Davson, H. and Rakić, L. J. (1985). Permeability of the blood–CSF and blood–brain barrier to thyrotropin releasing hormones. *Brain Res.*, **358**, 191–199
32. McComb, J. G., Davson, H. and Hollingworth, J. R. (1985). Further studies in the difference between ventricular and subarachnoid perfusion. *Brain Res.*, **89**, 81–91
33. Zlovokić, B. V., Davson, H., Preston, J. and Segal M. B. (1987). The effects of aluminium chloride on rate of secretion of the cerebrospinal fluid. *Exp. Neurol.*, **98**, 436–452

Recollections of Hugh Davson

William Oldendorf

First Meeting

I first met Hugh by accident. In the summer of 1964 I happened into a lecture hall at UCLA in which there was a lecture in progress. A scholarly looking and sounding Englishman was speaking on cerebrospinal fluid. I had heard that Hugh was in the States setting up the eye institute in Louisville. That the lecturer was indeed Hugh was confirmed at the end of his talk, when the host faculty member thanked him for his lecture. I then walked to the front of the hall, approached Hugh, introduced myself (he hadn't the least notion of who I was) and briefly told him of my interest in BBB. He listened politely, and when I asked if he could visit my laboratory he suggested we go immediately since his time was short. We went to my small lab and, after showing him my attempts at measuring human BBB *in vivo*, I mentioned that I had thought of visiting him in London in the fall, when I planned to enquire whether I might spend a year in his laboratory, beginning in the summer of 1965. To my astonishment, he offered me, on the spot, the opportunity of spending the proposed year with him. ('Why wait until the fall?') This was about two hours after I had wandered into his lecture and first laid eyes on him. With no formalities and no correspondence he signed me up. That year in his laboratory was a delight and completely redirected my research thinking.

Academic Honesty

Shortly after starting that year, the Royal Society of Medicine was presenting a one-day symposium on membrane permeability. Hugh was the first speaker and discussed the BBB. I attended and listened with interest but, of course, contributed nothing. About two years later the symposium appeared in print in the Society's proceedings. The first article was Hugh's material on BBB, but to my astonishment, I was second author. I rapidly scanned the article for a possible explanation. In two sentences Hugh speculated that the brain capillary was transformed into a

BBB by a humoral influence generated by glial cells. This was my suggestion during a discussion with Hugh at least two years before this article was published. Hugh had thought the speculation had merit and took the opportunity of getting it conveniently published in this Royal Society of Medicine symposium. A lesser person might have done so with no acknowledgement of its origin. In this case his acknowledgement was provided by putting me on as second author. I have watched with interest the many publications during the past 10–15 years which support the early speculation about a glial humoral role in BBB formation.

The Brain Uptake Index (BUI)

In 1969 I developed the BBB permeability measurement determining the fraction of a carbon-labelled test substance taken up by rat brain during a single microcirculatory pass following a rapid carotid injection. Tritiated water was injected in the same solution as a diffusible internal standard, with decapitation of the rat 5 seconds after injection. Our initial results suggested that it could be quite useful, and I prepared them for publication. Uncertain of the theoretical soundness and originality of the idea, I first sent the material to Hugh for his comment. He replied 'I have discussed your idea with Keasley (Welch) and we believe it to be sound and possibly valuable. In addition, it shows what a year in my laboratory can do for someone'.

The Devon Death March

In April, 1970, on a Sunday, I was scheduled to visit Hugh at his cottage in Devon. My train arrived at Barnstaple station at noon. Hugh and Marjorie were waiting. Hugh tossed my suitcase in the back of their car, whereupon Marjorie drove off. I was perplexed by this, but Hugh explained that it was such a pleasant day he thought we would walk to the cottage. He was a little vague about how far it was to the cottage, but surely 'only a few miles'.

We began our walk at about noon and for two or three hours a lively discussion ensued, but as the miles passed, it became clear at about 5 p.m. that we were lost. We had been so engrossed in our conversation that we had walked down some wrong road. There was about a mile of railway track down which we had walked which Hugh didn't recognize. At about 7 p.m. it was turning dark and I was becoming concerned. Much of this concern was due to the nature of the one-lane, high-walled roads which allowed only an occasional glimpse of the surrounding landscape for orientation. Hugh was especially concerned because we had been on the road for about seven hours and, being Sunday, the few pubs we had passed were closed. Our problem was suddenly resolved by the appearance in our path of an automobile driven by another house-guest of Hugh's, dispatched by Marjorie to scour the local roads for us. Actually we were only about half a mile from the cottage when rescued. Hugh later dubbed our protracted walk *The Devon Death March*.

1
Membrane Transport in Ocular Epithelia

J. A. Zadunaisky

INTRODUCTION

The study of transport across the ocular cell membranes has advanced substantially in recent decades. A historical approach to the developments in this field may give a perspective of the state of the art, especially when the knowledge available in the 1960s, when I first met Hugh Davson, is compared with the information we possess in the 1990s.

The rapid development of basic research in studies of the eye and vision, especially in the northern hemisphere in this period, made possible the study of solute and water transport mechanisms, and the discovery and presence of specific channels, cotransporters and exchangers. The research was encouraged by an audience avid for basic concepts to nourish the expanding clinical and surgical developments in ophthalmology.

The physiological phenomena known up to the 1960s had to be explained on the basis of cellular mechanisms and, especially in this field, mechanisms residing in the cell membranes. The pioneering work of Hugh Davson and his students in Great Britain as well as Kinsey and his associates in America had substantiated important facts. They demonstrated that three basic species were secreted into the aqueous humour: sodium chloride, bicarbonate and ascorbic acid. They also showed that corneal transparancy was an event dependent on metabolic energy and perhaps controlled by ionic pumps; they proved that the same principles applied to the crystalline lens and provided the ground work for the understanding of cataract formation. At that time the consensus was that the lens consisted of, or could be compared to, a giant cell, fibres and epithelium functioning as a single unit. The blood–ocular barriers were defined as located at the level of the ciliary body and the retinal pigment epithelium.

At this point it was clear that two lines of fundamental research had to be

"

pursued. One was the study of solute and fluid transport in a thermodynamically sound manner, as was happening in other areas of knowledge outside the eye. The second important route was the study of the pharmacology of the already established mechanisms. Under the influence of Hugh Davson, E. J. Conway, Hans Ussing, A. K. Solomon and Aaron Katchalski, to mention a few of the most relevant figures of the time in the field of membrane transport, we entered the field of eye research in order to study transport processes with the assurance that the association with Davson could give a band of young investigators. The purpose was then to describe and define the active and passive mechanisms by which solutes crossed ocular epithelia and their cell membranes. The methods used consisted of the isolation of tissues in the Ussing chamber, with the measurements of electrical potentials and short-circuit current aided by the determination of fluxes with radioisotopes. The detection of intracellular potentials with conventional and ion-sensitive microelectrodes and activities was used. Other techniques were developed later, such as the patch clamp method for the detection of membrane channels, the use of cell membrane vesicles and more recently the application of specific fluorescent probes to the study of ion movement. The expectation was and is that the dynamics of the movements of species across the membrane can be described electrophysiologically and optically and the mechanism of control or regulation be understood, as a prelude to possible application to

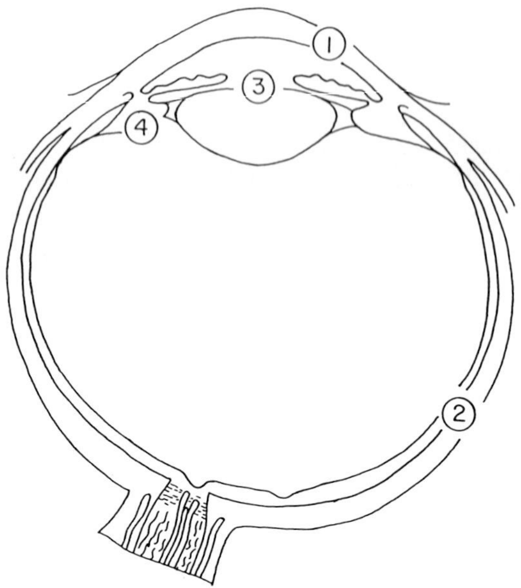

Figure 1.1   The epithelia of the eye. 1, Corneal epithelium and endothelium; 2, retinal pigment epithelium; 3, lens epithelium; and 4, ciliary epithelium

diseased states in ophthalmology. From the clear characterization of the kinetic events in the membrane, we know that the presence of specific molecular entities, such as channel proteins, could be predicted. From now on we must tackle the problem of isolation of these molecules. This last broad area will be a good subject to be summarized some 20 years from now and therefore only minimal reference to this budding field in membrane transport will be made here.

CORNEAL EPITHELIUM

The corneal electrical potential arises from its epithelium and is the consequence of both a net transport of chloride ions from aqueous to tear side in frog and toad and the combination of this same chloride transfer plus sodium active transport inward in mammalian species. The combination of chloride outward, sodium inward was first found in the corneal epithelium during stimulation of chloride secretion.

The active chloride transport of the corneal epithelium occurs as the combination of an active entry step at the basolateral side of the epithelial cell and exit through specific chloride channels located in the apical membrane. The model depicted in Figure 1.2 shows that the sodium dependence of Cl^- transport can be explained by the presence of a cotransporter moving two chloride ions, one sodium ion and one potassium ion into the epithelial cells. The 2ClNaK cotransporter utilizes the Na^+ gradient provided by the Na^+–K^+ pump to ensure the entry of Cl. The Cl concentration in the cell is 3–4 times greater than that expected from simple electrochemical equilibrium. The high Cl^- activity in the cell then overcomes the electrical gradient across the apical membrane and permits the passage of chloride through a conductance residing in the channels specifically permeable to the chloride ion. The channels are highly regulated by the sympathetic nervous system innervating the corneal epithelium. The addition of adrenaline to the epithelium produces a rapid increase in chloride secretion through the opening of the chloride channels. It has recently been shown that in the corneal epithelium a phosphorylation event is also needed for the opening of the channels.

The existence of chloride secondary active transport was first demonstrated in the corneal epithelium and since then in many other organs, including the intestine, the airways epithelia and kidney tubules. Activation by cathecholamines or agents that increase cyclic AMP in the cell was demonstrated simultaneously in the cornea, with a similar finding in the intestinal epithelium.

All hormonal, pharmacological agents, drugs or toxic agents that increase cyclic AMP stimulate chloride transport. The consensus is that activation of adenylate cyclase in the basolateral membranes produces an

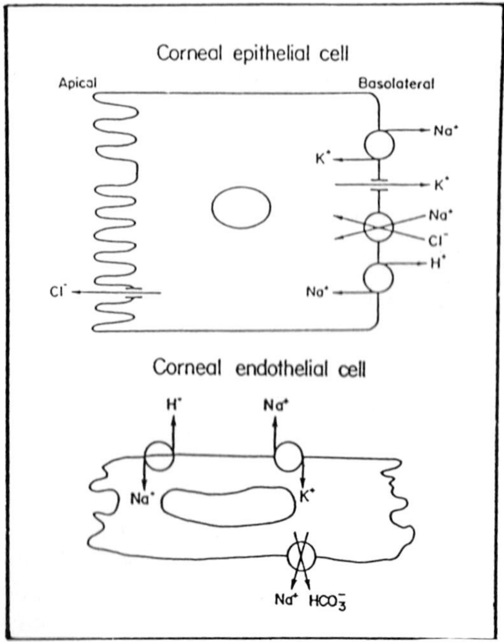

Figure 1.2   The limiting layers of the cornea. The epithelium pumps chloride outward, and in some species sodium inward simultaneously. The NaCl cotransporter depicted in the basolateral membrane of the corneal epithelium is most probably a 2ClNaK cotransporter, sensitive to loop diuretics. The corneal endothelium is responsible for the bulk of water movements across the cornea; the basic mechanism appears to be an Na bicarbonate transport from stroma to tear side

increase in cyclic AMP in the cell, which then activates the Cl⁻ channel through phosphorylation of cytokinase C in the apical membrane. The characteristics of the chloride channels in the corneal epithelium are similar to those found in other epithelia.

This chloride secretory system, as indicated, has now been found in numerous organs and tissues. Of particular interest is the case of the bronchial and tracheal epithelium. The main defect in the hereditary disease cystic fibrosis, affecting a large number of neonates, consists of a closure or lack of response of the chloride channels of the apical membrane to the activating effects of cyclic AMP. The recent discovery of the defective gene responsible for the disease has permitted the prediction of the conformation of the channel protein. Because of the similarity of the electrical properties of the channels found in the apical membranes of the cornea and of the trachea, it is very likely that the corneal apical membrane chloride channels consist of a protein of the type discovered in the lung epithelium.

The entry step for chloride is mediated by the 2ClNaK cotransporter of the basolateral membranes and inhibited by the loop diuretics furosemide and bumetamide. This inhibition reduces chloride activity in the cell and stops chloride secretion. In the case of the absorptive chloride uptake that occurs in the loop of Henle in the kidney, loop diuretics produce diuresis and are used in medical practice. The 2ClNaK cotransporter is activated by catecholamines and by cell shrinkage in nucleated red cells, and it could be possible that activation occurs by these agents also in the corneal epithelium.

The chloride channels can be blocked by specific agents, such as anthracenic acid and DNP. In our experience, the chloride secretion and the conductance of the apical membrane channels measured with intracellular microelectrodes produced only partial inhibition, of the order of 50% of the initial chloride conductance. The chloride channel blockers developed and tested by Greger produce an increase in the chloride activity in the epithelial cells.

As indicated in Figure 1.2, the intracellular pH of the epithelial cells is regulated by the presence of an Na^+/H^+ exchanger that is activated when the intracellular pH is displaced from its normal level. Recent experimentation confirms the presence in these cells of a pH regulatory system. Indirect evidence appears to show that a Cl^-/HCO_3 system could also be in operation in the corneal epithelial cells. K^+ channels have been characterized in the basolateral membrane of the corneal epithelial cells, by examination of the effects of Ba^{2+}.

RETINAL PIGMENT EPITHELIUM

The retinal pigment epithelium (RPE) was mounted in an Ussing-type chamber for ion studies by Lasanski and De Fisch in 1966. These authors showed the presence of chloride active transport from the apical to the basolateral side, as well as sodium transport in the opposite direction. They observed also that a fraction of the current was extremely sensitive to the concentration of bicarbonate in the medium bathing the preparation. Steinberg and Miller in 1970 took up this preparation and produced a series of papers that opened up this field and clarified the understanding of the ways in which this membrane performs as a barrier and controls the microenvironment of the photoreceptors. The mechanisms currently thought to be present in the cell membranes of the RPE are shown in Figure 1.3. This tissue, together with the choroid plexus, shows higher concentrations of Na^+K^+ ATPase in the apical villous membrane that intimately embraces the outer segments of the photoreceptors or the retina. Several basic mechanisms in the membranes of the RPE have now been confirmed in a number of species. The Na^+K^+ pump is located in the apical membranes and transports sodium inward into the subretinal space

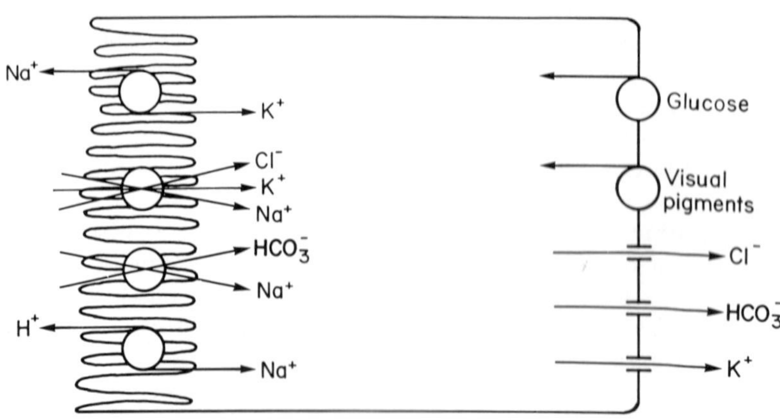

Figure 1.3   Pumps, exchangers, cotransporters and channels so far described in the retinal pigment epithelium. The data were obtained in several species as well as in tissue cultures of retinal pigment epithelium and should be considered as a composite of expected membrane events of this epithelium

and K into the cell. The low sodium in the cell facilitates the entry of Cl$^-$ via a 2ClNaK cotransporter also located in the apical membrane. This cotransporter has similar characteristics to those of that described above for the corneal epithelium: it transports chloride into the cell, it is inhibited by loop diuretics and it is activated by cyclic AMP. The electrophysiological model described in our laboratory predicted the existence of Cl$^-$ channels in the vasolateral membrane which permit the exit of Cl$^-$ to the choroidal side of the tissue.

The original work of the 1960s demonstrated a dependence on bicarbonate of the current circulating through the isolated RPE. This aspect has been examined in our laboratory and in the laboratory of Sheldon Miller.

The bicarbonate dependence indicates the existence of specific mechanisms for the control of intracellular pH. The presence of an Na$^+$/H$^+$ exchanger was demonstrated by the use of membrane vesicles of the apical membrane of the RPE. The creation of a pH gradient between the inside of the vesicles and the medium induces an increase in the Na uptake. This increase in Na$^+$ uptake is inhibited by amiloride at high concentrations and is not affected by agents that disrupt electrochemical gradients such as valinomicyn, which produces cationic channels or depolarization with high concentrations of K$^+$. The presence of the Na$^+$/H$^+$ exchanger has thus far been found in shark and bovine RPE.

The other component of the intracellular pH regulation which a bicarbonate dependence suggests is the possible presence of a Cl$^-$/HCO$_3^-$ exchanger. In fact, it has been demonstrated that such an exchanger is present in the basolateral membrane of intact RPE of the frog, in Miller's laboratory.

The consequences of these two systems, an Na^+/H^+ exchanger and a Cl^-/HCO_3^- exchanger in the basolateral membranes, is of importance for the conservation and recovery of the normal pH of the subretinal space. During light activation of the photoreceptors, a rapid alkalinization of the subretinal space has been demonstrated recently by Steinberg. The phenomenon is reversible, and it is suggested that the Na^+/H^+ and Cl^-/HCO_3^- exchangers of the RPE are responsible for the return of the subretinal pH to normal levels. Together with the glial-like properties of the RPE, i.e. a high K^+ permeability, the picture develops of a specialized cell interposed between the outer retinal and the choroidal vessels which regulates the ionic and pH environment of the photoreceptors.

The passage of sugars through the RPE has been re-examined in both intact animals and *in vitro* preparations. The initial contention that there was an active transport of D-glucose into the subretinal space has not been substantiated. There is entry of D-glucose in preference to L-glucose, but by a facilitated diffusion mechanism and not by active net transport of sugars. In intact rats the entry of glucose can be increased in advanced stages of diabetes and retinal degeneration. However, during the initial stages the entry is not affected and it can be concluded that lack of or reduced sugar transport is not a causative factor in these two models of retinal diseases, advanced streptozotozin diabetes and spontaneous degeneration in the RCS rat. The defect observed in advanced stages indicates an opening of the blood retinal barrier to both the non-transportable L-glucose and the transported D-glucose. The active component of D-glucose entry appears to be unaffected even when the barrier is partially opened by the degenerative process. These studies on the barriers to glucose have elicited an interest in the entry and distribution of ascorbic acid, partially motivated by the possible function of ascorbic acid as a protecting agent during the action of free radicals.

CRYSTALLINE LENS

The original concept of the lens as a unit, or large cell, transporting Na^+ out and K^+ inward through the pump of the epithelium, useful as it was in its time, has given way to the more appropriate concept of the fibres and epithelium as autonomic units, linked through the existence of multiple gap junctions between them. In the first place the fibres of the lens are normal cells with all the attributes of selectivity and metabolic activities. They are interconnected as a syncitium by gap junctions and this gives the lens its unitary characteristics. Rae and Mathias have done much to illumine this point by utilizing conductance measurements in a three-dimensional mode, using fine microelectrodes inserted into lens fibres and recording extra- or intracellularly at distant points. The resulting picture is one of homogeneous spread of currents and permitted the conclusions expressed above. Changes in experimental conditions, including pH manipulations, can uncouple the junctions and modify the syncitial nature in a predictable manner.

The lens epithelium, the site of the Na^+/K^+ pump, helps keep all the organs at an intracellular low Na^+ concentration, facilitating the existence of cotransporters and exchange mechanisms that utilize the sodium gradient. The fibres have all the necessary machinery to perform as cells, but at a reduced metabolic rate in comparison with the epithelium.

The question of lens transparency is linked to the integrity of the fibre membranes. It is possible that the initial defect in ageing leading to the formation of cataracts is a change in the permeability or a derangement of a pump, channels or exchanger in the fibre membrane. This change could induce an imbalance intracellularly in the fibre which could result in general ionic changes. In turn, this could lead to the formation of clusters of proteins contained in the fibres, the crystallines, which will scatter light. When the clusters reach a size larger than one-quarter of the wavelength of the incident light, they scatter light and result in the opacification of the lens.

The use of the patch clamp technique for the detection of ion channels has permitted the finding of numerous channels on both the epithelium and the fibres of the lens: K^+ channels, Na^+ channels and Cl^- channels. In fact, a great variety of ionic channels have been found and the relevance of each of them for the function of the epithelium and fibres is being worked out at present. The numerous events observed with this powerful technique in the lens and in other tissues has now to be integrated into the general physiological function of the specific tissues or organs.

More recently, the use of fibre and epithelial membrane vesicles has permitted the characterization of membrane events, shown in Figure 1.4.

An Na^+/H^+ exchanger in both the epithelium and the fibres is involved in control of intracellular pH. The 2ClNaK cotransporter is also present as well as the Na^+/Ca^{2+} exchanger described by Galvan and now found also by us with the aid of fluorescent probes. A Ca^{2+} ATPase has also been found in isolated vesicles from lens fibres and the complexity of the membrane of the fibres can only be emphasized by these findings.

Sugar enters the lens through a transport mechanism requiring Na. In this sense the carrier is very similar to that described for the intestinal mucosa. A study utilizing membrane fragments of the lens indicates that the molecular configuration of the sugar carrier in the lens is probably very similar to that found and extensively studied elsewhere in experimental animals.

The process of sorbitol accumulation in the lens during development of

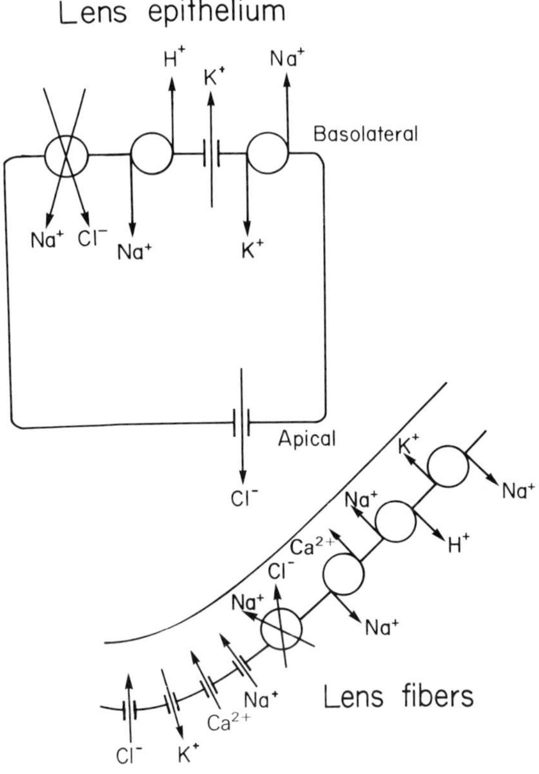

Figure 1.4　Mechanisms located on the cell membranes of the lens epithelium and lens fibres. A number of new membrane events have been described recently in membranes of the lens fibres indicating more and more their behaviour as active cellular elements

diabetes, attributed to the retention of the alcohol in the fibres inducing water retention osmotically, has produced an immense literature, including reports on the search for inhibitors of aldose reductase, the enzyme responsible for the conversion of sugars to sorbitol. This concept, which for years in the work of Kinoshita (1965) and his group appeared to be confined to the lens, its opacification and the development of cataracts in diabetics, has been extended to other organs and falls within the area of osmotic regulations of cell water, volume and transport. In fact, the accumulation of sorbitol has been found in tissues other than the lens in diabetics, including peripheral nerves. The reaction of the cell to changes in osmotic activity in the plasma, in this case due to rises and falls in the uncontrolled levels of D-glucose in the blood and extracellular space, produces a defence reaction in cells in general which falls into the category of regulatory volume increase and regulatory volume decrease. Cells, and therefore lens fibres, tend to accumulate osmotically active agents when the osmolarity increases in the extracelllular medium. In the case of diabetes, the mechanism appears to be the induction of the formation of sorbitol to compensate for the high sugar level in the aqueous and vitreous humour.

Oxidative agents have been studied extensively because of their tendency to induce opacification of the lens and the nature of high production and slow removal of free radical from the lens. However, little has been done to link the action of these agents to the mechanism in the lens fibre relating to ion movements. It is possible that control of intracellular pH is deranged during the noxious action of oxidating agents. Recent investigations with isolated vesicles from peripheral lens fibres appear to indicate an activation of the existing Na^+/H^+ exchanger in the fibre membranes by oxidating agents.

CILIARY EPITHELIUM

In the last decade substantial advances have occurred in the study of the mechanism of formation of the aqueous humour by the ciliary epithelium. The basic knowledge of transport of sodium chloride, bicarbonate and ascorbic acid has been augmented by the location of pumps, channels and exchangers in the appropriate membranes. The ciliary epithelium has represented a serious challenge to epitheliologists involved in solving secretory mechanisms in the eye, because of its anatomical configuration, the delicate characteristics of the tissue and the unsuspected nature of the relationship of the two epithelia, the pigmented and the non-pigmented. The attempts of David Cole in the early 1960s proved their value in the 1980s, when several groups were able to produce preparations that physiologically permit the partial understanding of the formation of the fluid. Although more research is required to give a complete picture of the process controlling intraocular pressure, the secretory process at the level of the ciliary epithelium appears now less of an insoluble problem, thanks to the use of a combination of electrophysiological experiments and tissue cultures.

Krupin and Candia took up the preparation of the rabbit iris ciliary body and developed the first reliable *in vitro* preparation in a modified Ussing chamber. Wiederholt and Zadunaisky utilized the wide band of shark's eye ciliary epithelium for the same purpose.

The most significant demonstration was that both the pigmented and the non-pigmented layers were involved in the process of formation of the fluid. The resting potentials and currents in this preparation are sensitive to the classical inhibitors and stimulators of transport. However, there is a different response when the agents are placed on the pigmented or non-pigmented sides of the tissue. The usual behaviour of epithelia is that agents or drugs with high specificity will act only on one side of the tissue and, if they can diffuse across it, will always show either inhibition or stimulation. In the case of the complex ciliary epithelia, it is found that the response can be stimulation in one side and inhibition in the other. The rabbit ciliary body showed sodium transport and an inverse response to ouabain on each side. The preparation developed by Wiederholt and Zadunaisky utilizing the 4-mm-wide ciliary body of the shark, mounted without the iris, also showed the same paradoxical responses. Furthermore, the activation of the electrophysiological parameters could occur as a result of the action of agents that increase cyclic AMP in the cells, such as forskolin, pointing to the existence of Cl^- movements. The investigation with microelectrodes also showed that the two cells, pigmented and non-pigmented, are at isopotential levels. The lucid descriptions of Guiseppina Raviola of the anatomy of the ciliary epithelium permitted the definition of gap junctions between the two cells and can explain the fact that both cells are at isopotential difference. That is, the two cells communicate and the ionic equilibrium is the same in both of them at least as regards the Cl^- ion, tested with specific Cl^--sensitive microelectrodes. Although the rabbit does not show similar properties, it is a fact that the ciliary epithelium shows great species differences and investigators are forced to deduce the function of these layers as a composite of the information obtained in different animals. Cl^- secretion has been known to exist in the ciliary epithelium of the cat since 1959, evidence being provided by the experiments of Monte Holland.

The other significant development in this area has been the use of tissue cultures. The preparation of cell lines from the pigmented and the non-pigmented epithelium in tissue cultures by Coca-Prados has given an enormous impetus to the field. The unselfish distribution of these cell lines to physiological investigators has permitted the recording of several events in the membranes of both types of ciliary epithelium.

The group of Wiederholt in Berlin demonstrated the presence of a number of unsuspected Na^+-dependent cotransporters and exchangers in the membranes of the pigmented epithelium. Although the cells in culture lose their polarity and therefore it is difficult to ascertain whether the membrane proteins showing electrical responses are polarized, the combination of the data from the cultures and the intact preparation has permitted the development of the model shown in Figure 1.5.

Another development has also been the use of the split ciliary epithelium. Recently, in our laboratory and in collaboration with the group of anatomists led by Elke Drukell-Lutjen, we have observed that split preparations retain electrical and transport properties of the pigmented layer, confirming effects that were found in the intact preparations. Furthermore, the existence of gap junctions between the two cell layers has been confirmed also in the shark ciliary epithelia. The model shown in

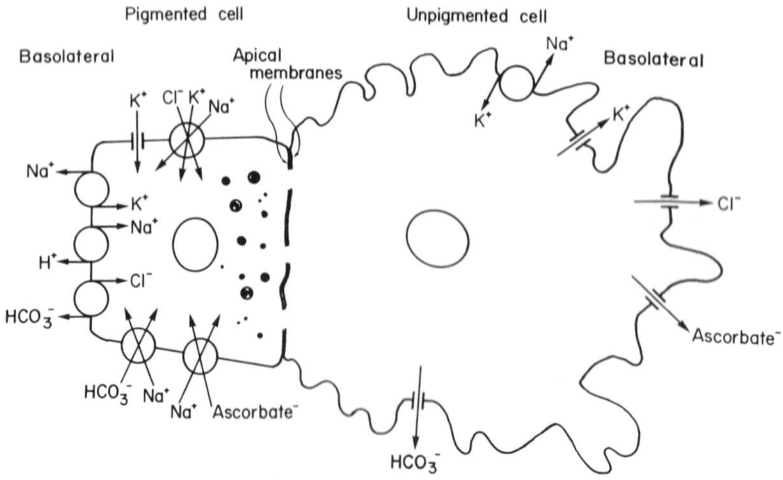

Figure 1.5 A modern model for the ciliary epithelia; both cells, pigmented and non-pigmented, are involved in the secretion and formation of the aqueous humour

Figure 1.5 can be better understood if the two layers are compared with one cell only. The pigmented layer basolateral membrane, facing the blood side, contains the entry step for a number of ion movements, including an Na^+K^+ pump, a 2ClNaK cotransporter, a Cl^-/HCO_3^- exchanger, an $Na^+HCO^-_3$ cotransporter, an Na^+/H^+ exchanger and an Na^+ ascorbate$^-$ cotransporter. This will produce a low level of Na in the two cells, because of the free passage through the gap junctions, a high level of K, a high ascorbate content and control of the intracellular pH. The properties of the opposing apical membranes have been difficult to assess, but it is thought that the presence of the gap junctions will nullify gradients of concentration between the two cells. The basolateral membrane of the non-pigmented epithelium will contain the appropriate channels for the exit of the three important species that are transported into the posterior chamber and induce the movement of water for the formation of the aqueous humour. The channels of the basolateral membrane of the non-pigmented layer will then contain a chloride channel, a sodium channel, a bicarbonate channel and an ascorbate channel.

CONCLUSION

The model presented here is the best approximation to the complexity of the system and it is possible that further investigations will permit modifications that will stress the importance of some of the mechanisms or perhaps their location in the membranes. However, the general criteria of two cell layers opposed by their apical membranes, with gap junctions and one cell acting as an apical membrane and the other as a basolateral one, is perhaps the best development of recent years in the area of aqueous humour formation.

BIBLIOGRAPHY

Davson, H. (1970). *A Textbook of General Physiology*. Churchill, London

Davson, H. (1990). *The Physiology of the Eye*, 5th edn. Macmillan Press, London

Di Mattio, J. and Zadunaisky, J. A. (1986). Facilitated glucose transport across the retinal pigment epithelia of the bull frog (*Rana catesbana*). *Exp. Eye Res.*, **43**, 15–28

Holland, M. G. and Gipsance, C. C. (1970). Chloride transport in the isolated ciliary body. *Invest. Ophthalmol.*, **9**, 20–29

Kinoshita, J. H. (1965). Pathways of glucose metabolism in the lens. *Invest. Ophthalmol.*, **4**, 619–628

Krupin, T., Reinach, P. S., Candia, O. A. and Podos, S. M. (1984). Transepithelial electrical measurement on the isolated iris/ciliary body. *Exp. Eye Res.*, **38**, 115–123

Lansanski, A. and De Fisch, F. W. (1966). Potential current and ionic fluxes across pigment epithelium and choroid. *J. Gen. Physiol.*, **49**, 913–914

Miller, S. S. and Steinberg, R. H. (1977). Active transport of ions across frog retinal pigment epithelium. *Exp. Eye Res.*, **25**, 235–240

Pascuzzo, G. J., Johnson, J. E. and Pautler, E. L. (1980). Glucose transport in the isolated mammalian pigment epithelium. *Exp. Eye Res.*, **30**, 53–56

Rae, J. L. (1974). Potential profiles in the crystalline lens of the frog. *Exp. Eye Res.*, **19**, 235–242

Raviola, G. (1977). The structural basis of the blood–ocular barrier. *Exp. Eye Res.*, **25** (Suppl.), 27–63

Steinberg, R. H. and Miller, S. (1973). Aspects of electrolyte transport in frog pigment epithelia. *Exp. Eye Res.*, **22**, 639–645

Zadunaisky, J. A., Lande, M. A. and Hafner, J. (1971). Further studies on chloride transport in the frog cornea. *Am. J. Physiol.*, **221**, 1832–1836

2

Transport Functions of the Blood–Ocular and Blood–Brain Barriers, and the Microenvironment of Neuronal and Non-neuronal Tissues

Laszlo Z. Bito

INTRODUCTION

The origin and nature of intraocular and cerebrospinal fluids (IOFs and CSFs) has been of particular interest to a relatively small, but highly motivated, group of scientists. After considerable growing pains, these fields of inquiry have made great strides over the last few decades, largely due to Hugh Davson, to whom this volume is dedicated.

In this contribution, I shall focus on the current state of understanding of the nature, origin and chemical composition of IOFs. In keeping with the scope of this volume, I shall emphasize some of the similarities and differences between these fluids, especially as they relate to the unique requirements of the different tissues of the eye, as opposed to the less diverse nature of the central nervous tissues, whose extracellular fluids are in a diffusional equilibrium with CSF.

Inadequate analytical and sampling techniques had led to the incorrect conclusions that there are no substantial differences between the composition of blood plasma and of IOFs, with the notable exception of a much lower protein concentration in the latter, and that aqueous humour represents a filtrate of plasma (Duke-Elder, 1932). These views prevailed until some investigators, among them most notably Hugh Davson, began to employ more refined analytical techniques and recognized the need to obtain samples under steady-state conditions.

THE EXISTENCE OF ACTIVE TRANSPORT PROCESSES ACROSS THE BLOOD–OCULAR BARRIERS

The most elegant demonstration that substances that can neither be produced nor destroyed within the eye are not distributed between blood and aqueous humour according to a passive Donnan equilibrium, was based on the determination of sodium and chloride in the aqueous humour and blood plasma of dogs and cats before and after these fluids were dialysed against each other (Davson *et al.*, 1949). These experiments have demonstrated that the aqueous humour of these species had an excess of NaCl as compared with that of a plasma ultrafiltrate and have proved most elegantly that active secretory processes must play a role in the formation of aqueous humour (see Davson, 1956, 1969, 1984).

Further studies using more and more refined analytical techniques suggested that a great number of substances show a distribution that is inconsistent with unmodified ultrafiltration (see Cole, 1984; Bito, 1985). Ascorbic acid is of particular interest in this respect, since its concentration in some species, including the human, is some 20- to 30-fold greater in aqueous humour than in blood plasma (Cole, 1984; Bito, 1985).

Studies with radioactively labelled substances, such as amino acids and glucose, or their non-metabolizable analogues, revealed that even crystalloids that are not present in an excess in aqueous humour may also cross these barriers by a carrier-mediated, saturable process. Such transport processes from blood into the aqueous humour and active removal of amino acids across the retinal region of the blood–ocular barriers maintain relatively high amino acid concentration in the anterior segment and very low levels of amino acids at the retina (Reddy *et al.*, 1961; Bito *et al.*, 1966).

SIMILARITIES AND DIFFERENCES BETWEEN AQUEOUS HUMOUR AND CEREBROSPINAL FLUID

The apparent similarities between CSF and aqueous humour were celebrated by the classic treatise entitled *Physiology of Ocular and Cerebrospinal Fluid*, one of the first among the many outstanding contributions of Hugh Davson (1956) to the scientific literature.

Since then, even more exacting studies, especially those allowing direct comparison of IOFs and CSFs taken from different regions of these fluid systems, revealed that the aqueous humour is uniquely different from CSF and that the greatest similarity between CSF and IOF is typically found in the posterior regions of the vitreous compartment adjacent to the neuro-retina (Bito and Davson, 1964; Bito, 1970; Bito and DeRousseau, 1980; Cole, 1984).

THE NATURE OF THE VITREOUS BODY AND THE FLUID CONTAINED WITHIN IT

In human and some other primates, the vitreous body is a gel at birth, but a liquid pocket develops in its centre soon thereafter, due to the withdrawal of collagen fibres from that region, as the globe enlarges (Balazs and Denlinger, 1984). This liquid pocket increases in size as the collagen network further collapses after it loses its central support.

Thus, the vitreous body is by no means a homogeneous entity. In fact, it may be best visualized as a liquid compartment surrounded by an unevenly distributed network of collagen fibrils and, at least in the young, interspersed with collagen fibres providing a gel state. However, the collagenous network can be easily removed by simple filtration (Bito and Davson, 1964).

We can, therefore, talk about vitreous humour—that is, the fluid contained within the fibrous network of the vitreous body—as we talk about other extracellular fluids, including fluid contained between the collagen fibres of connective tissues. We can also count the vitreous body among the IOFs, because it is 98% water, its gel structure being achieved with collagen fibrils representing only about 1% of its total weight. The remaining 1% represents electrolytes and other solutes, most of which are typically present in all biological fluids, although not necessarily at the same concentrations. As we shall see below, physiologically highly significant concentration gradients are maintained across this compartment with respect to many solutes to serve the greatly different requirements of the neuroretina as opposed to the peripheral-type tissues of the anterior chamber.

The gel nature of the vitreous in most primates, or its highly viscous nature in others, must be of paramount importance in this regard, because such required concentration differences could not be maintained across a low-viscosity-fluid volume that would be subject to mixing caused by thermal gradients, rapid eye movements and pulsation. Thus, in addition to possible advantages with regard to protection of the retina from physical damage (see Balazs and Denlinger, 1984), the gel nature of the vitreous body must be essential to maintain the CNS-like chemical environment required by the neuroretina, as opposed to the peripheral type of environment required by tissues of the anterior segment. Thus, any age-dependent deterioration of this gel, or its destruction by vitrectomy, can be expected to lead to a deterioration, or peripheralization, of the chemical environment of the retina and, hence, may contribute to retinal stress and pathologies.

CONCENTRATION GRADIENTS WITHIN AND BETWEEN INTRAOCULAR FLUID COMPARTMENTS

Early attempts at defining concentration gradients within the IOFs frequently used dissection of the frozen globe. However, this technique was found to lead to artefacts due to the freezing out of solutes. We have, therefore, developed a Plexiglas platform over which the unfrozen globe could be held and bisected in its equatorial plane with a microtome knife; this blade becomes embedded in a rubber pad, preventing the fluid released from the two separated halves of the vitreous body from intermixing before sampling (Bito and Davson, 1964).

By use of this technique, it was shown, for example, that the concentrations of K^+ and amino acids are higher in the anterior as compared with the posterior half of the vitreous (Bito and Davson, 1964; Bita *et al.*, 1966), while the opposite gradient has been observed with respect to Mg^{2+} (Bito, 1970). Since the vitreous body does not effectively restrict the diffusion of small molecules (Maurice, 1957), this implies that there is a continuous influx of K^+ and amino acids into the anterior segment of the eye through the ciliary epithelia, and an active removal of these solutes against a concentration gradient back into blood across the blood–retinal barrier system. Mg^{2+}, on the other hand, is actively secreted from blood into the retina, presumably by both the retinal capillaries and the choroidal epithelium (Bito, 1970; Bito and DeRousseau, 1980). Thus, the neuro-retina has a low K^+, low amino acid and high Mg^{2+} extracellular environment similar to, and possibly identical with, that of the cerebral cortex (Bito and Davson, 1966; Bito, 1969).

THE RESERVOIR FUNCTION OF THE VITREOUS BODY AND THE PROTECTION OF NEURAL TISSUES DURING ACUTE ASPHYXIA

Because of its very large volume, as compared with that of the retina, the vitreous compartment can be regarded as an effective emergency reservoir that can minimize the accumulation in the retinal extracellular fluid of such solutes as lactic acid and K^+, while it can serve as a potential source of glucose during asphyxia or circulatory arrest (Bito and DeRousseau, 1980). This is in contrast to the very thin film of CSF covering the cerebral cortex, allowing the development of very rapid and drastic changes in the composition of extracellular environment of cortical cells during acute asphyxia (Bito and Myers, 1972).

It has been argued that such rapid changes, including the precipitous rise in extracellular K^+, leads to the shut-down of the cerebral cortex, while limitations on cellular swelling prevent permanent damage to cortical cells during episodes of acute asphyxia. According to this hypothesis, the cortex is more vulnerable to prolonged partial asphyxia than to acute, complete asphyxia, while other regions of the brain, whose shut-down is incompatible with survival, could obviously not have evolved a similar shut-down mechanism and are only protected against acute asphyxia by redundancy (Bito and Myers, 1972). It is possible that the very large volume of the vitreous reservoir available to the retina—possibly combined with some degree of hindrance of cellular swelling because of the unique physical properties of the vitreous, especially of its cortical layer adjacent to the retina—represents yet another evolutionary mechanism to protect highly vulnerable neuronal tissues against acute asphyxia.

Finally, the vitreous compartment must be regarded as part of the lymphatic system of the retina, which, like the brain, lacks conventional lymphatic drainage systems (cf. Bito and DeRousseau, 1980).

THE BLOOD–OCULAR BARRIERS AND THEIR TRANSPORT FUNCTIONS

The barrier systems of the eye are analogous to the well-known barrier systems of the brain, except that the former appears to be even more complex than the latter. We can come to this conclusion since, as we have seen, the transport functions existing across the blood–ocular barrier system have to maintain two different chemical environments—a peripheral type in the anterior segment and a CNS type in the posterior segment of the globe, in the vicinity of the retina. Furthermore, these different environments must be maintained in the absence of a strict diffusional barrier between the anterior and posterior segments, entirely by different transport functions associated with different regions of the ocular barrier system (Bito and DeRousseau, 1980).

It must be noted that even though tissues of the anterior segment, as opposed to the retina, do not seem to require an overall chemical environment uniquely different from that of most other peripheral tissues of the body, the maintenance of tight-junctional barriers between blood and aqueous humour is of paramount importance, since this fluid must be essentially protein-free to allow optical clarity. Such a barrier must have effective transport mechanism for solutes that are required by these tissues, in order to allow the passage of these solutes at sufficient rates into the continuously flowing aqueous humour (Bito and DeRousseau, 1980).

By the same token, once such barriers have evolved, the removal of all but the most lipid-soluble substances that are produced within the eye and the brain require special transport mechanisms. These can be termed the 'absorptive transport systems of the blood–tissue barriers' as opposed to the 'secretory transport functions' of these same barrier systems (Bito and

Wallenstein, 1977; Bito and DeRousseau, 1980) and can prevent the passage of autacoids from the retina into the anterior chamber of the eye (see below; see also Bito and Wallenstein, 1977).

The location of what we must regard to be the most sophisticated transporting tissue of the body, namely the ciliary processes, is of strategic importance in this regard, since these processes project into the narrow posterior chamber, which lies between the anterior chamber and the vitreous body. Thus, the ciliary processes are optimally situated within the eye to physiologically separate, in the absence of an anatomical separation, the different environments required by the peripheral type of tissues of the anterior and the neuronal tissue of the posterior segment of the eye (Bito and DeRousseau, 1980), and can prevent the passage of autacoids from the retina into the anterior chamber of the eye (see below; see also Bito and Wallenstein, 1977).

THE ABSORPTIVE TRANSPORT SYSTEMS OF THE BLOOD–OCULAR AND BLOOD–BRAIN BARRIERS

It has been known for several decades that both the ciliary processes of the eye and the choroid plexuses of the brain can actively accumulate certain inorganic anions and organic acids *in vitro*, when incubated in the presence of these substances (Becker, 1960; Barany, 1972; Bito *et al.*, 1976a) and can remove these substances from the IOFs and CSFs. The combined use of ventriculocisternal and corticocisternal perfusions allowed the characterization of such absorptive transport process (Bito *et al.*, 1966) and this combined perfusion technique eventually led to the resolution of one of the most controversial aspects of neurophysiology, i.e. the true size of the extracellular compartment of the brain, which was grossly underestimated in size, or even refused recognition of existence, by early electronmicroscopists.

However, the test substances that were used for studying this absorptive transport process such as para-aminohippuric acid, iodide and various X-ray contrast media, such as iodipamide (Barany, 1972), were substrates that are not produced within the eye and the brain, do not readily penetrate these organs from the circulation and, for the most part, were unlikely to be present in the circulation throughout evolutionary development. Thus, the reason for the evolution of these absorptive transport processes represented a challenge to physiologists (Cole, 1970).

However, more recent studies revealed that these transport processes are likely to have evolved to remove from the extracellular compartments of the eye and the brain autacoids, such as prostaglandins (PGs), that are produced within these organs. The need for absorptive transport for some autacoids that are not locally metabolized and can not pass through lipid barriers is obvious, since their accumulation within the eye or the brain would clearly interfere with normal physiological functions (Bito and Wallenstein, 1977; Bito, 1986, 1987).

THE ABSORPTIVE TRANSPORT OF PROSTAGLANDINS ACROSS THE BLOOD–OCULAR AND BLOOD–BRAIN BARRIERS

The nature, physiological role, specificity and kinetic parameters of the PG transport system of the brain, the eye and other organ systems (lung, kidney, uterus) have been reviewed repeatedly. These reviews also discuss the vulnerability of these transport systems to damage and the possible role of the failure of these transport systems in pathological processes (Bito and Wallenstein, 1977; Bito, 1986, 1987). For these reasons, only their salient features are stated here.

PGs are produced within the eye and the brain normally, and in excessive amounts during pathological processes. While these tissues may modify PGs, they do not completely metabolize or inactivate them. Since the basic cell membrane is impermeable to these autacoids, PGs that are produced within the eye and the brain must be removed by a carrier-mediated process across the tight-junctional blood–ocular and blood–brain barriers. Such transport processes have indeed been demonstrated, both *in vivo* and *in vitro*.

The isolated ciliary body and the choroid plexus of all species studied so far were found to actively accumulate PGs against a concentration gradient by a saturable mechanism that is also competitively inhibited by probenecid, a classical inhibitor of the organic acid transport system. *In vivo* studies have demonstrated that the rate of removal of intravitreally injected $PGF_{2\alpha}$ is much more rapid than the rate of loss of sucrose, a non-transported substance of virtually identical molecular weight (Bito and Salvador, 1972). The carrier-mediated removal of $PGF_{2\alpha}$ from CSF has been demonstrated by ventriculocisternal perfusion studies (Bito *et al.*, 1976).

PATHOPHYSIOLOGICAL AND PATHOLOGICAL CONSEQUENCES OF THE LOSS OR INHIBITION OF PG TRANSPORT

Inhibition of this PG transport system was shown to enhance the deleterious effects of intravitreal PG injection on the retina and also of supracortically applied PGs on cortical function (Bito and Wallenstein, 1977; Wallenstein and Bito, 1977). In fact, pretreatment with inhibitors of the organic acid transport system, such as probenecid and bromcresol green, allowed the induction of prostaglandin-induced epileptiform cortical activity and damage to the PG transport function of the BBB appears to be involved with the formation of latent epileptogenic foci (Bito and Wallenstein, 1977; Wallenstein and Bito, 1977). Surprisingly, inhibitors of PG transport were found to allow even the development of a hyperthermic response to a supracortically applied PG (Bito and Wallenstein, 1977).

While the neurobiology community has not taken much notice of the potential physiological significance of these PG transport systems, the above studies clearly suggest that these mechanisms are essential for normal function. Thus, their loss or temporary inhibition can be deleterious. These are important considerations, since many drugs have already been shown to be effective inhibitors of PG transport (Bito and Salvador, 1976; Bito, 1987). Interestingly, indomethacin, the classical inhibitor of PG synthesis, was shown to stimulate PG transport in low concentrations, and to inhibit it at high concentrations.

A better understanding of the nature of these transport processes, and the effects of various drugs on the rate of removal of PGs from the eye and the brain, can be expected to provide a new approach to the understanding of the pathophysiology of some ocular and neuronal disorders. Furthermore, a better understanding of these absorptive transport systems can be expected to provide a means to manipulate the distribution of endogenous eicosanoids, and their removal from the extracellular fluids of the brain and the eye, as well as their delivery to these organs for therapeutic purposes (Bito, 1987).

It should be mentioned in this regard that, in contrast to earlier studies that assigned to PGs a deleterious or even pathological role in the eye, more recent studies clearly demonstrate that, with the notable exception of rabbits, PGs have important physiological roles in the eye. In fact, PGs have great therapeutic potential with respect to the control of ocular inflammation and especially with regard to the medical management of glaucoma (Bito, 1984; Bito and Stjernschantz, 1989). Thus, a better understanding of the pharmacokinetics of such endogenous autacoids, and drugs derived from such autacoids, can be expected to provide new approaches to the management of ocular, neurological and psychiatric disorders.

REFERENCES

Balazs, E. A. and Denlinger, J. (1984). The vitreous. In Davson, H. (Ed.), *The Eye*, Vol. IA. Academic Press, London, pp. 533–589

Barany, E. H. (1972). Inhibition by hippurate and probenecid of *in vitro* uptake of iodipamide and *o*-iodohippurate. A composite uptake system for iodipamide in choroid plexus, kidney cortex and anterior uvea of several species. *Acta Physiol. Scand.*, **86**, 12–27

Becker, B. (1960). The transport of organic anions by the rabbit eye. I. *In vitro* iodipyracet (diodrast) accumulation by ciliary body–iris preparations. *Am. J. Ophthalmol.*, **50**, 862–867

Bito, L. Z. (1963). Blood–brain barrier: Evidence for active cation transport between blood and the extracellular fluid of brain. *Science*, **165**, 81–83

Bito, L. Z. (1970). Intraocular fluid dynamics. I. Steady-state concentration gradients of magnesium, potassium and calcium in relation to the sites and mechanisms of ocular cation transport processes. *Exp. Eye Res.*, **10**, 102–116

Bito, L. Z. (1984). Prostaglandins, other eicosanoids, and their derivatives as potential antiglaucoma agents. In Drance, S. M. and Neufeld, A. H. (Eds), *Glaucoma: Applied Pharmacology in Medical Treatment*. Grune and Stratton, New York, pp. 477–506

Bito, L. Z. (1985). Composition of intraocular fluids and the microenvironment of the retina. In Lajtha, A. (Ed.), *Handbook of Neurochemistry*, Vol. 8. Plenum Press, New York, pp. 231–252

Bito, L. Z. (1986). Absorptive transport of prostaglandins and other eicosanoids across the blood–brain barrier system and its physiological significance. In Suckling, A. J., Rumsby, M. G. and Bradbury, M. W. B. (Eds), *The Blood–Brain Barrier in Health and Disease*. Ellis Horwood, Chichester, pp. 109–121

Bito, L. Z. (1987). Eicosanoid transport systems: Mechanisms, physiological roles, and inhibitors. In Willis, A. L. (Ed.), *Handbook of Eicosanoids*, Vol. 1. CRC Press, Boca Raton, Florida, pp. 145–160

Bito, L. Z., Bradbury, M. W. B. and Davson, H. (1966). Factors affecting the distribution of iodide and bromide in the central nervous system. *J. Physiol. (London)*, **185**, 323–354

Bito, L. Z. and Davson, H. (1964). Steady-state concentrations of potassium in the ocular fluids. *Exp. Eye Res.*, **3**, 283–297

Bito, L. Z., Davson, H. and Fenstermacher, J. (Eds) (1977). *The Ocular and Cerebrospinal Fluids. Fogarty International Symposium*. Academic Press, London

Bito, L. Z., Davson, H. and Hollingsworth, J. R. (1976a). Facilitated transport of prostaglandins across the blood–cerebrospinal fluid and blood–brain barriers. *J. Physiol. (London)*, **256**, 273–285

Bito, L. Z., Davson, H., Levin, E., Murray, M. and Snider, N. (1966). The concentrations of free amino acids and other electrolytes in cerebrospinal fluid, *in vivo* dialysate of brain, and blood plasma of the dog. *J. Neurochem.*, **13**, 1057–1067

Bito, L. Z., Davson, H. and Salvador, E. V. (1976b). Inhibition of *in vitro* concentrative prostaglandin accumulation by prostaglandins, prostaglandin analogues and by some inhibitors of organic anion transport. *J. Physiol. (London)*, **256**, 257–271

Bito, L. Z. and DeRousseau, C. J. (1980). Transport functions of the blood–retinal barrier system and the micro-environment of the retina. In Cunha-Vaz, J. G. (Ed.), *The Blood–Retinal Barriers*. Plenum Press, New York, pp. 133–163

Bito, L. Z. and Myers, R. E. (1972). On the physiological response of the cerebral cortex to acute stress (reversible asphyxia). *J. Physiol. (London)*, **221**, 349–370

Bito, L. Z. and Salvador, E. V. (1972). Intraocular fluid dynamics. III. The site and mechanism of prostaglandin transfer across the blood intraocular fluid barriers. *Exp. Eye Res.*, **14**, 233–241

Bito, L. Z. and Salvador, E. V. (1976). Effects of anti-inflammatory agents and some other drugs on prostaglandin biotransport. *J. Pharmacol. Exp. Ther.*, **198**, 481–488

Bito, L. Z. and Stjernschantz, J. (Eds) (1989). *The Ocular Effects of Prostaglandins and Other Eicosanoids*. Alan R. Liss, New York

Bito, L. Z. and Wallenstein, M. C. (1977). Transport of prostaglandins across the blood–brain and blood–aqueous barriers and the physiological significance of these absorptive transport processes. In Bito, L. Z., Davson, H. and Fenstermacher, J. (Eds), *The Ocular and Cerebrospinal Fluids. Fogarty International Symposium*. Academic Press, London, pp. 229–243

Cole, D. F. (1970). Aqueous and ciliary body. In Graymore, C. N. (Ed.), *Biochemistry of the Eye*. Academic Press, London, pp. 105–181

Cole, D. F. (1984). Ocular fluids. In Davson, H. (Ed.), *The Eye*, Vol. IA. Academic Press, London, pp. 269–390

Davson, H. (1956). *Physiology of the Ocular and Cerebrospinal Fluids*. Churchill, London

Davson H. (Ed.) (1969). *The Eye*, Vol. I: *Vegetative Physiology and Biochemistry*. Academic Press, London

Davson, H. (Ed.) (1984). *The Eye* (3rd edn), Vol. IA: *Vegetative Physiology and Biochemistry*. Academic Press, London

Davson, H., Duke-Elder, W. S. and Maurice, D. M. (1949). Changes in ionic distribution following dialysis of aqueous humour against plasma. *J. Physiol. (London)*, **109**, 32–40

Duke-Elder, W. S. (Ed.) (1932). *A Textbook of Ophthalmology*, Vol. I. Kimpton, London

Maurice, D. M. (1957). The exchange of sodium between the vitreous body and the blood and aqueous humour. *J. Physiol. (London)*, **137**, 110–125

Reddy, D. V. N., Rosenberg, C. and Kinsey, V. E. (1961). Steady-state distribution of free amino acids in the aqueous humor, vitreous body and plasma of the rabbit. *Exp. Eye Res.*, **1**, 175–181

Wallenstein, M. C. and Bito, L. Z. (1977). Prostaglandin E_1-induced alterations in visually evoked response and production of epileptiform activity. *Neuropharmacology*, **16**, 687–694

3

Ocular Pulsatile Blood Flow in Healthy and Diabetic Eyes

Maurice E. Langham

I am delighted to join with so many friends and admirers of Hugh Davson to pay tribute on this very happy occasion. Hugh is recognized throughout the world as one of the select and august group of English physiologists who contributed numerous fundamental contributions to the Medical Sciences in the early and middle decades of the twentieth century.

Hugh had already made outstanding contributions to membrane physiology when I first took courage to seek a research position in his laboratory in the Department of Physiology, University College, London. I had just completed an honours degree in Physiology under Sir Charles Lovatt-Evans and his exceptional faculty, which included A. V. Hill, Bernard Katz and L. E. Bayliss. It was with trepidation that I knocked on the door and walked into Hugh's laboratory. The scene was one I would come to know very well in the following months and years: Hugh seated comfortably at his desk, relaxed, legs outstretched, puffing quietly on his pipe. His thoughts were far away and indeed it took several moments before he noticed my arrival; perhaps he was thinking about his ongoing animal experiments, for there on the floor with heads protruding from the wooden boxes were a group of rabbits patiently awaiting his bidding. I had no need to be fearful; Hugh emerged from his thoughts with his characteristic smile and immediately put me at ease. A few words to me and I was asked when I would like to start work—tomorrow or next week? It was his relaxed attitude, friendly and sincere manner, covering an intense deep commitment to his research, that I was privileged to enjoy in the months and years to come.

The daily teatime breaks, with tea served by his faithful secretary, Molly Kirk, became an enjoyable ritual and gave the young investigators precious moments to seek his advice. It was a sad day for us at The Institute of Ophthalmology when he returned to University College.

The research interest that Hugh stimulated in me on the character and control of fluid circulation in the eye has continued throughout my career. Our hope, shared by Hugh, has been to increase understanding of the physiological mechanisms determining the stability of the intraocular fluid circulation; thereby to provide the basis for new therapies for the prevention and treatment of ischaemic diseases of the eye, the principal causes of world-wide blindness. Therefore, it seems appropriate to review some of the progress recently made towards this goal.

The flow of blood into the eye is pulsatile and causes a rhythmic fluctuation of the steady-state intraocular pressure (IOP). It is the amplitude and the form of this pulse (Figure 3.1) that is used to calculate the rate of pulsatile blood flow (PBF) into the eye. The simple fluid model depicted in Figure 3.2 illustrates the fundamental physical concept needed to translate pressure measurements into flow determinations. Discrete quantities, or boluses, of incompressible fluid flow through an elastic chamber. When a bolus enters the chamber, its volume is accommodated by the chamber walls, which stretch and exert an increased pressure on the fluid. Clearly, if this expansion and contraction occurs f times per second and the volume of

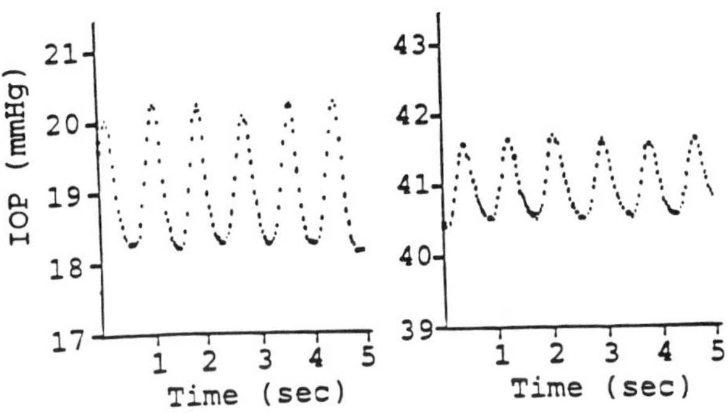

Figure 3.1   IOP pulse recordings in a normal eye. The measurements were made at intervals of 30 ms. The second set of readings were made when the IOP had been increased by use of a scleral suction cup

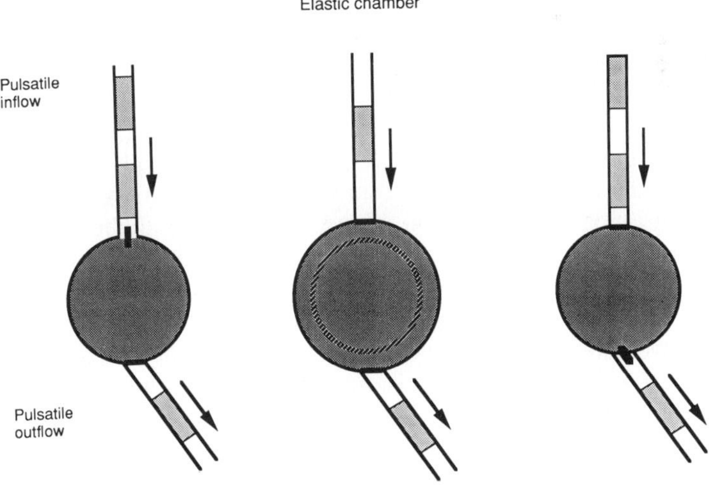

Figure 3.2   Fluid model to illustrate the pulsatile mechanism of ocular blood flow into an elastic chamber (the eye); the simple picture involves a bolus inflow accompanied by volume expansion, followed later by a bolus outflow and volume contraction

an individual bolus is v_0, then the flow is fv_0. Since f is measured directly, the challenge is to obtain v_0. Because the volume is so small, it would be extremely difficult to measure the dimensions of the chamber with sufficient accuracy to obtain volume change. However, if there is an independent determination of the relationship between volume and pressure, pressure measurements can be used to obtain the volume of a bolus and thereby to determine the flow.

The simple bolus model of Figure 3.2 would require valves in order to have the inflow and outflow occur at separate times. However, a more realistic picture of the eye would include steady outflow at all times, with a pulsatile inflow superimposed. Figure 3.3 gives an example of the application of this process to the tonometric measurements of the IOP (Figure 3.3a). The pressure–volume relationship is used to convert the IOP readings into intraocular volume changes vs time (Figure 3.3b). Differentiating the latter curve with respect to time yields the instantaneous component of the net PBF (Figure 3.3c). The average rate of PBF inflow is then calculated by the application of mathematics with continuity relations and closure to back-integrate the flow curves to get the volume change curve (see Langham *et al.*, 1989).

The PBF in eyes of healthy young adult subjects is of the order of 700–900 μl/min (Table 3.1). The PBF is almost exclusively ciliary choroidal flow and far exceeds the total retinal blood flow of 33 ± 9.6 μl/min (Riva *et al.*, 1985) to 80 ± 13 μl/min (Feke *et al.*, 1989a) measured by laser Doppler velocimetry (Riva and Feke, 1981). The PBF decreases smoothly with increased IOP in a non-linear manner and is zero at an IOP equal to the ophthalmic arterial systolic pressure (Langham and To'mey, 1978; Langham *et al.*, 1981; Langham and Preziosi, 1984).

In diabetic eyes the PBF has been found to decrease proportionately with the severity of the retinopathy (Table 3.2). Also, in diabetic patients with moderate to severe retinopathy the ophthalmic arterial pressures were lower than in diabetic patients with minimal retinopathy (Table 3.2).

The finding that choroidal blood flow may be abnormally low in the eyes with diabetic retinopathy is consistent with the observations of Feke *et al.* (1989b) of a decreased retinal blood flow in the diabetic eye. The substantial decrease of the choroidal pulsatile blood flow with increasing severity of the retinopathy indicates a corresponding increase of vascular resistance (Feke *et al.*, 1989a). This, however, contrasts with the view of the absence of morphological changes in the choroid in patients with early diabetes (Ashton, 1969; Kohner *et al.*, 1969). The alternative explanation is that the decrease of choroidal blood flow results from localized vasoconstriction. In this respect the natural potent vasoconstrictor peptide angiotensin 2 is known to modulate the systemic circulation in diabetes.

Angiotensin 1 is a decapeptide formed from the action of renin on its substrate, angiotensinogen. It has no known biological action in humans,

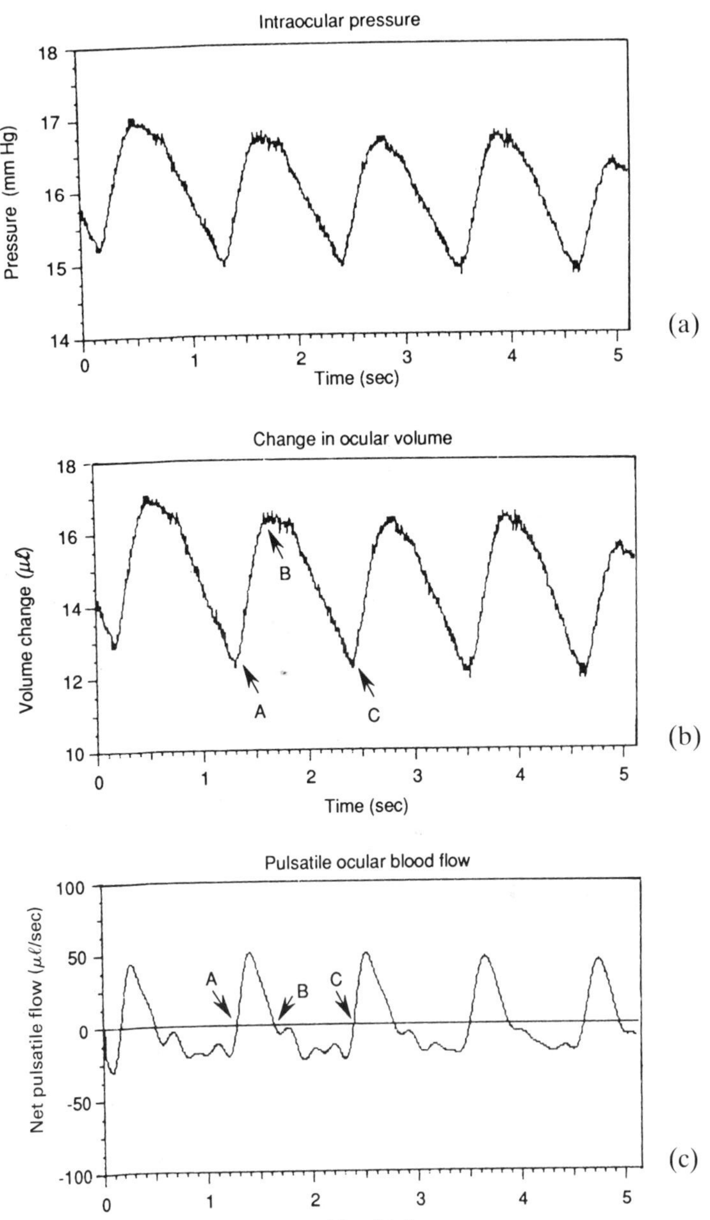

Figure 3.3 The net pulsatile ocular blood flow in a normal healthy subject. The top figure (3.3a) shows the instantaneous IOP measurements taken at 10 ms intervals. The middle figure (3.3b) gives the change in ocular volume relative to a reference ocular volume at 10 mmHg. The data correspond to the IOP measurements of the upper figure. The lower figure (3.3c) gives the net pulsatile ocular blood flow corresponding to the above figures

                    Barriers and Fluids of the Eye and Brain

Table 3.1   Ocular haemodynamics in healthy eyes of adult subjects

Parameter	Right eye	Left eye	Difference
Pulsatile blood flow (μl/min)	724 $\pm$42 (8)	718 $\pm$37 (8)	6.6 $\pm$6.5 (8)
Diastolic IOP (mmHg)	15.5$\pm$ 1.0 (8)	15.6$\pm$ 0.8 (8)	0.43$\pm$0.2 (8)
Systolic IOP (mmHg)	17.9$\pm$ 1.0 (8)	18.1$\pm$ 0.9 (8)	0.45$\pm$0.2 (8)
Ophthalmic arterial pressure (OAP) (mmHg)	87$\pm$2		—
Diastolic brachial arterial pressure (mmHg)	73$\pm$4 (8)		—
Systolic brachial arterial pressure (mmHg)	138$\pm$3		—

The results are expressed in terms of the arithmetic mean $\pm$ the standard error of the mean; the numbers of eyes studied are given in parentheses. The difference column is based on measurements on pairs of eyes of individual subjects. The ophthalmic pressures were equal in pairs of eyes and the values are expressed as the mean for the 10 patients.

Table 3.2   The minimal (diastolic) and maximal (systolic) IOPs and the PBF in eyes of normal and diabetic subjects. Group 1 are diabetics with no apparent retinopathy; Group 2 are diabetics with background retino-pathy; Group 3 are diabetics with proliferative retinopathy

Group	N	IOP (mmHg)	PA (mmHg)	PBF (μl/min)
Normals	19	15.5$\pm$1.0 17.9$\pm$1.0	2.3$\pm$0.1	648$\pm$42
Group 1	9	16.3$\pm$0.7 18.5$\pm$0.7	2.1$\pm$0.2	570$\pm$32
Group 2	11	17.2$\pm$0.6 19.1$\pm$0.6	1.5$\pm$0.3	471$\pm$70
Group 3	13	16.7$\pm$0.7 17.5$\pm$0.7	0.8$\pm$0.1	210$\pm$37

The results are expressed as the arithmetic mean $\pm$ the SE of the mean using mean values from pairs of eyes (values were equal in pairs of eyes); N is the number of patients in the group.

but is rapidly converted by angiotensin-converting enzyme (ACE) to the highly active vasoconstrictor octapeptide, angiotensin 2. ACE has been identified in ocular tissues (Strittmatter *et al.*, 1989) and angiotensin has been shown to constrict ocular vessels (Rockwood *et al.*, 1987). A number of inhibitors of ACE have been used clinically and found to decrease peripheral vascular resistance and systemic blood pressure (see Oates and Wood, 1990; William, 1990). Inhibition of ACE has been observed to block the constrictor action of angiotensin on retinal vessels (Rockwood *et al.*, 1987). Treatment of hypertensive and diabetic patients with ACE inhibitors increased the PBF (Schilder, Hopkins and Langham, unpublished). In other clinical studies deterioration of renal function and proteinuria in diabetic patients decreased when treated orally with ACE inhibitors (Bjork *et al.*, 1986; Hommel *et al.*, 1986).

In conclusion, a non-invasive technique now exists for the rapid measurement of the ocular pulsatile blood flow. This has allowed assessment of the ciliary choroidal blood flow in normal and diseased eyes. In diabetics the choroidal pulsatile blood flow decreases with the severity of the retinopathy. Consequently, not only is it necessary to evaluate the retinal pathology in the diabetics but, perhaps more importantly, it is necessary to know how the pulsatile blood flow is affected.

REFERENCES

Ashton, N. (1969). In P. Almaric. Albi (Ed.), *Fluorescein Angiography*. Karger, Basel, pp. 334–345

Bjork, S., Nyberg, G., Mulec, H., Granerus, G., Herlitz, H. and Aurell, M. (1986). Beneficial effects of angiotensin converting enzyme inhibition on renal function in insulin-dependent diabetic patients with nephrology. *Br. J. Med.*, **295**, 471–474

Feke, G., Bazney, N., Goger, D., Stock, N. and Gabbay, K. (1989a). Variation of retinal blood flow with duration of disease in type 1 diabetes. *ARVO*. Sarasota, Florida

Feke, G., Goger, D., Sebag, J. and Weiter, J. (1989b). Blood flow in the normal retina. *Invest. Ophthalmol. Vis. Sci.*, **30**, 58–65

Hommel, E., Parving, H., Mathiesen, E., Edsberg, B., Damkjaer-Nielson, M. and Giese, J. (1986). Effect of captopril on kidney function in insulin-dependent diabetic patients with nephropathy. *Br. J. Med.*, **293**, 467–470

Kohner, E., Dollery, C. and Bulpitt, C. (1969). Cotton-wool spots in diabetic retinopathy. *Diabetes*, **18**, 691–704

Langham, M., Farrell, R., O'Brien, V., Silver, D. and Schilder, P. (1989). Blood flow in the human eye. *Acta Ophthalmol.*, **191**, 9–13

Langham, M. and Preziosi, T. (1984). Non-invasive diagnosis of mild to severe stenosis of the internal carotid artery. *Stroke*, **15** (4), 614–620

Langham, M. and To'mey, K. (1978). A clinical procedure for the measurement of the ocular pulse–pressure relationship and the ophthalmic arterial pressure. *Exp. Eye Res.*, **27**, 17–25

Langham, M., To'mey, K. and Preziosi, T. (1981). Carotid occlusive disease: The effect of complete occlusion of the internal carotid artery on the intraocular pulse/pressure relation and on the ophthalmic arterial pressure. *Stroke*, **12**, 759–765

Riva, C. and Feke, C. (1981). Laser doppler velocimetry in measurement of retinal blood flow. In Goldman, H. (Ed.), *The Biomedical Laser: Technology and Clinical Application*. Springer-Verlag, New York, pp. 135–161

Riva, C., Grunwald, J., Sinclair, S. and Petrig, B. (1985). Blood velocity and volumetric flow rate in human retinal vessels. *Invest. Ophthalmol. Vis. Sci.*, **26**, 1124–1132

Rockwood, E., Fantes, F., Davis, E. and Anderson, D. R. (1987). The response of retinal vasculature to angiotensin. *Invest. Ophthalmol. Vis. Sci.*, **28**, 677–683

Strittmatter, S., Braas, K. and Snyder, S. (1989). Localization of angiotensin converting enzyme in the ciliary epithelium of the rat eye. *Invest. Ophthalmol. Vis. Sci.*, **30**, 2209–2214

Williams, G. H. (1990). Converting enzyme inhibitors in the treatment of hypertension. *New Engl. J. Med.*, **23**, 1517–1525

4
Physiology of the Vitreous: A Personal View

D. M. Maurice

If I remember correctly, Maurice Langham and I joined Hugh on the same day in 1945, this surely making us the earliest of his still active students. For many years many of our colleagues, especially in the USA, believed Maurice and I were one person; perhaps they still do. My project at that time involved the vitreous humour and the first papers (Davson *et al.*, 1949; Duke-Elder *et al.*, 1949) on which my name appeared were on this topic. I have since written (Maurice, 1980): 'The vitreous body, corresponding in function to the space between the lens and the film of a camera, is very properly ignored by most sensible people.' This gibe was aimed at Dr E. A. Balazs, but on preparing this chapter I was bemused to find that my entanglement with the jelly was more than a youthful indiscretion, since I have authored a further ten full-length papers on its properties, and several others have been partially concerned with it. What made it interesting was that in spite of its apparent inactivity it repeatedly presented unexpected phenomena.

In the original studies with Hugh Davson we determined the rate of transport of tracers from the blood to the vitreous in order to characterize this barrier between them. This approach encountered two problems. First, it was limited to those tracers which would pass the barrier readily or could be detected at very low levels. Second, it could not easily distinguish between entry directly from the blood across the retinal surface or indirectly by way of the aqueous humour. I had noted a delayed rise in the aqueous/plasma ratio of ^{24}Na after its systemic injection and had correctly assigned it to the influence of the vitreous body (Maurice, 1951), an effect that was given elegant mathematical formulation by Friedenwald and Becker (1955).

To evade these problems, I injected the tracer directly into the vitreous body and followed its disappearance from the eye with an external Geiger counter. This disappearance was found to be logarithmic (Maurice, 1957) (Figure 4.1). The amount leaving by way of the anterior segment was estimated by determining the level of ^{24}Na in the aqueous humour, and

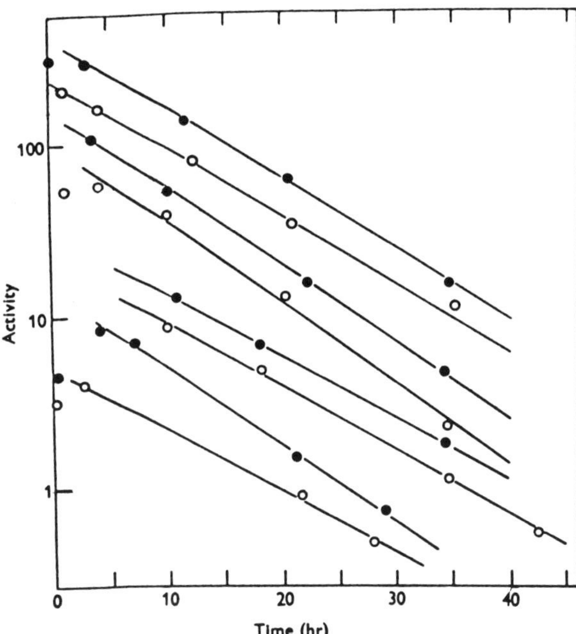

Figure 4.1 Rate of loss of ^{24}Na from eye after injection into vitreous humour, measured with external counter. Each group of points from one animal

amounted to about 40% of the total loss (Figure 4.2). The kinetics were compatible with free diffusion of the ion in a stagnant vitreous humour.

From there I used the same methods to examine the loss of radio-iodinated serum albumin (Maurice, 1959), which might be expected not to pass across the retinal surface but to leave the vitreous entirely by way of the anterior chamber. The level of activity in the aqueous humour corresponded to this hypothesis, which predicts that

$$C_a F = C_v k_v V_v \tag{4.1}$$

where C_a and C_v are the average concentrations of tracer in the aqueous and vitreous, f is the aqueous outflow rate, k_v is the loss constant from the vitreous and V_v is its volume. All these values can be measured in the experiment except f, which must be established by other techniques.

I modelled the diffusion of albumin within the vitreous by using a thermal analogue, the only occasion on which this has been attempted in biology, to the best of my knowledge. A 20 cm representation of a segment of the vitreous cavity was cast in lead and the surface corresponding to the anterior hyaloid was cooled with a jet of water at room temperature (Maurice, 1959) (Figure 4.3). The casting was then heated with a flame and its logarithmic drop in temperature was followed. The model was cali-

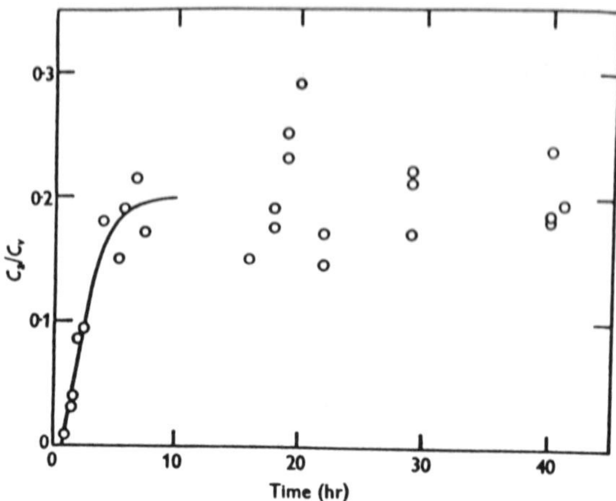

Figure 4.2 Rise in concentration of ^{24}Na in aqueous humour after its injection into vitreous body. Each point from one animal

brated on a rectangular brick of lead where the rate of temperature drop could be calculated theoretically, and it appeared to behave very well. The problem can be solved with a lot of effort on a computer (Moseley, 1981), but the analogue was more entertaining, if less versatile.

The results made it possible to calculate the theoretical rate of loss of a substance from the vitreous by the anterior route from a knowledge of its diffusion constant in the vitreous body. The molecular weight can be used to replace the diffusion constant, since they are strongly connected. This relationship and the relationship between k_v and C_a/C_v from Equation (4.1) are plotted in Figure 4.4 together with the experimental results for a variety of substances and it can be seen that there is a fair accordance for many of them (Maurice, 1980). On the other hand, Becker (Forbes and Becker, 1960; Becker, 1961) had shown that the behaviour of iodide and

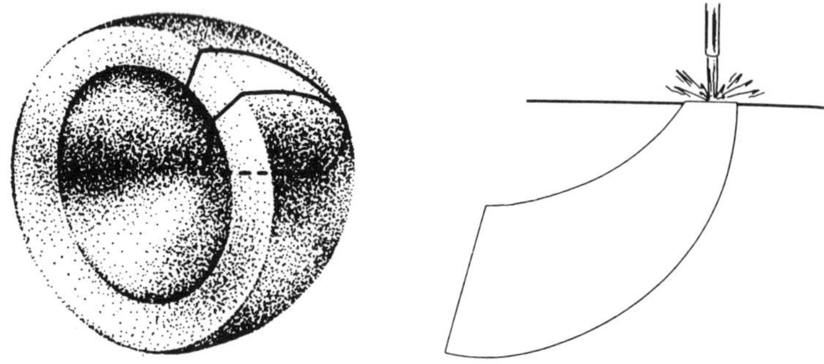

Figure 4.3 (*Left*) Outline of segment of vitreous body used to construct lead model for thermal analogue. (*Right*) Method of conducting experiment. Cold water is hosed against equivalent of anterior vitreous face of heated lead model. Rubber dam protects other surfaces

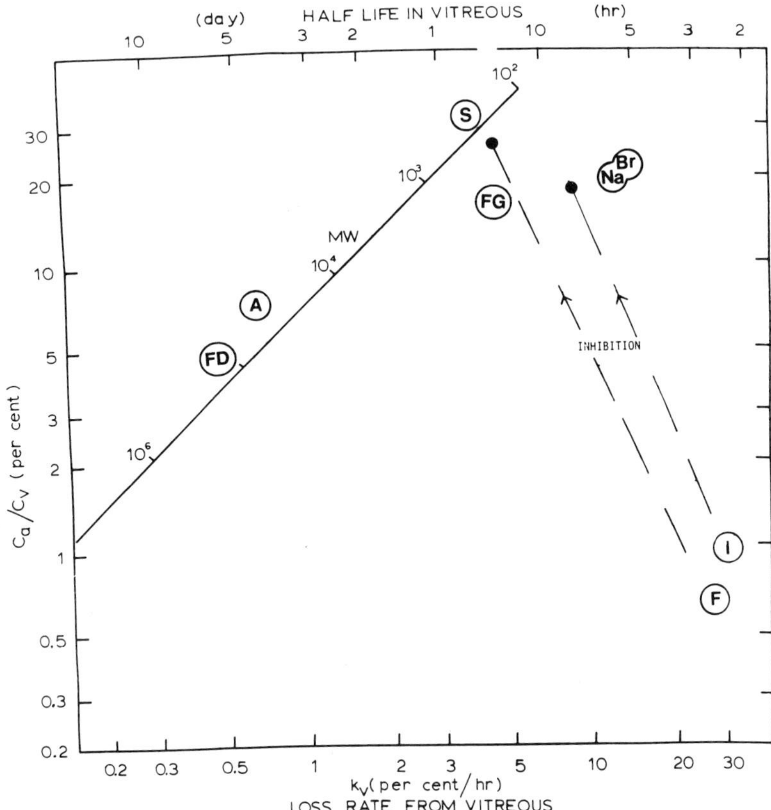

Figure 4.4　Line indicating theoretical relationship between the concentration ratio between aqueous and vitreous humour and the loss rate of tracer from eye. Molecular weight calibration was derived from thermal analogue. Experimental values for various tracers are indicated (Maurice, 1980; Araie and Maurice, 1989). S, sucrose; FG, fluorescein glucuronide; A, serum albumin; FD, FITC dextran, MW 70 000

iodopyracet was completely different; after injection into the vitreous, they left the eye far more rapidly and their C_a/C_v values were much smaller than would have been expected from the theoretical relationships. However, they became close to the theoretical when inhibitors were applied or saturating concentrations of the tracer were injected, indicating that they were being removed from the eye by an active process.

Together with Dr Cunha-Vaz, I established similar relationships for fluorescein. I have recounted elsewhere how this collaboration came into being (Maurice, 1985). It was happy in being distinguished by a complete regional division of authority: so he had primary responsibility for all problems on the retinal side, and I for those on the vitreous side of their boundary. An advantage of using a fluorescent compound is that the gradients of concentration within the vitreous body can be observed *in vivo* as well as *in vitro* and from this the diffusional fluxes of the dye across the boundaries of the body can be established (Figure 4.5). This technique made it clear that fluorescein was being lost from the vitreous across the retinal surface rather than by the anterior route (Cunha-Vaz and Maurice, 1967; Araie and Maurice, 1989). Further experiments showed that its transport was an active process and that the mechanisms were located both in the cells of the pigment epithelium and in those of the retinal capillary endothelium.

Fluorescein, in fact, penetrates very poorly into the vitreous from the blood both in the human and the rabbit eye. Under the enthusiastic guidance of Cunha-Vaz (1985) the determination of this penetration has come into clinical use, since its rise is a measure of the breakdown of the blood–vitreous barrier in certain diseases of the retina. As far as I know, there is as yet no direct evidence that fluorescein is actively transported out of the human eye and it suffers from being converted into its glucuronide in the blood as well as being slightly toxic. Its use in preference to better alternatives is a historical accident, since it was, presumably, the only fluorescent compound available to Ehrlich (Johnson and Maurice, 1984) in his pioneer experiments: it is no longer an easy matter to test other candidate dyes on human subjects.

The finding that solutes of large molecular weight diffuse slowly from the vitreous into the aqueous compartment suggested a method (Ehrlich, 1950) of determining the rate of aqueous flow and its fluctuations, in rabbits. A small quantity of fluorescein-labelled dextran was injected into the vitreous body and time was allowed for it to set up a steady state with the anterior chamber. After this, non-invasive fluorometric measurements of C_a should allow variations of the flow rate, f, to be determined according to Equation (4.1). The vitreous compartment served as a reservoir for the

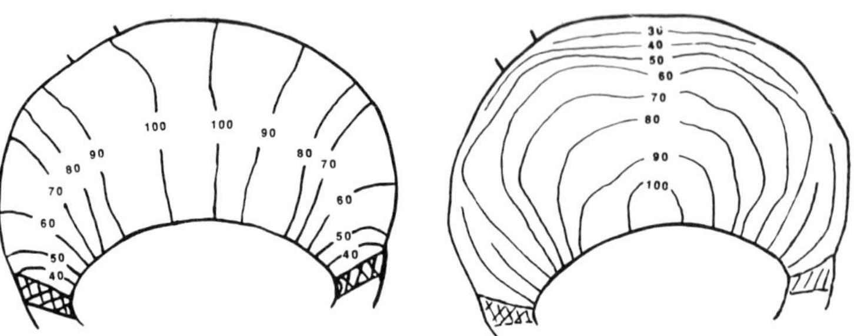

Figure 4.5 Vitreous concentration contours of tracers at steady state. (*Left*) Fluorescein glucuronide: concentration gradients indicate loss of the tracer principally into the anterior segment. (*Right*) Fluorescein: concentration gradients indicate loss of tracer principally across retina. Figures compiled from Araie and Maurice (1989)

dextran, which was released at virtually constant rate into the anterior chamber. This technique appeared to be very successful until an experiment was made in which cholera toxin was injected through the sclera into the vitreous, whereupon an enormous change in the aqueous level was noted. On carrying out a control experiment in which saline was injected, or even a needle hole was made, a similar larger change in the aqueous level was noted (Maurice, 1987). The explanation could be only that fluid was leaking out of the needle hole and being replaced by aqueous humour seeping back across the aqueous–vitreous interface, thus holding back the forward diffusion of dextran. A variety of subsidiary tests gave results compatible with this explanation. It was evident, then, that the level of fluorescent dextran in the anterior chamber was more sensitive to aqueous flow across the vitreous interface than to its bulk outflow from the anterior chamber. The results obtained by the technique have, therefore, to be interpreted with caution, particularly when a short-term change associated with a change in intraocular pressure is concerned.

Another conclusion that can be drawn from the experiments on the leakage of fluid out of a scleral needle hole is that there is normally very little seepage of aqueous humour backward across the vitreous interface. This is interesting because subretinal fluid is rapidly absorbed across the pigment epithelium, apparently by an active transport system. It is difficult to believe that this fluid volume could be replaced in the vitreous other than from the ciliary body, so that if the active system was transporting fluid at the same rate, under normal circumstances a marked anterior seepage should be detectable. It is piquant that one of the active investigators of the pigment epithelial transport system is Dr Marmor of the Ophthalmology Department of Stanford; we have not yet devised an experiment that will resolve the apparent contradiction.

Currently I am examining particulate markers which may be able to reveal small fluid drifts within the vitreous body. India ink has been frequently used to examine outflow pathways in the vitreous, but it seemed to me that too large quantities have been injected, which would obscure any subtle patterns of movement. We have had more success by injecting very small volumes of the ink, but even so it is difficult to visually track the drift of a diffuse cloud of greyish particles (Chopra and Maurice, 1988). Currently, other modalities, scintigraphy and MRI, are under investigation. I am convinced that the vitreous still has mysteries to reveal.

REFERENCES

Araie, M. and Maurice, D. M. (1989). The loss of fluorescein, fluorescein glucuronide, and FITC dextran from the vitreous. *Exp. Eye Res.* (in press)
Becker, B. (1961). Iodide transport by the rabbit eye. *Am. J. Physiol.*, **200**, 804

Chopra, A. and Maurice, D. (1988). Diffusion of colloid in the vitreous. *Proc. Int. Soc. Eye Res.*, **5**, 67

Cunha-Vaz, J. G. (1985). Vitreous fluorophotometry. In Osborne, N. N. and Chader, G. J. (Eds), *Progress in Retinal Research*. Pergamon Press, Oxford, pp. 89–114

Cunha-Vaz, J. G. and Maurice, D. M. (1967). The active transport of fluorescein by the retinal vessels and the retina. *J. Physiol. (London)*, **191**, 467–486

Davson, H., Duke-Elder, W. S., Maurice, D. M., Ross, E. J. and Woodin, A. M. (1949). The penetration of some electrolytes and non-electrolytes into the aqueous humour and vitreous body of the cat. *J. Physiol. (London)*, **108**, 203–217

Duke-Elder, Sir S., Davson, H. and Maurice, D. M. (1949). Studies on the intra-ocular fluids. 2. The penetration of certain ions into the aqueous humour and vitreous body. *Br. J. Ophthalmol.*, **33**, 329–338

Ehrlich, P. (1950). Contributions to the theory and practice of histological staining. Inaug. Diss. Leipzig. In *The Collected Papers of Paul Ehrlich*, Vol. 1. Pergamon Press, London

Forbes, M. and Becker, B. (1960). The transport of organic anions by the rabbit eye. II. *In vivo* transport of iodopyracet (Diodrast). *Am. J. Ophthalmol.*, **50**, 867

Friedenwald, J. S. and Becker, B. (1955). Aqueous humor dynamics. *Arch. Ophthalmol.*, **54**, 799–815

Johnson, F. and Maurice, D. (1984). A simple method of measuring aqueous humor flow with intravitreal fluoresceinated dextrans. *Exp. Eye Res.*, **39**, 791–805

Maurice, D. M. (1951). The permeability to sodium ions of the living rabbit's cornea. *J. Physiol. (London)*, **112**, 367–391

Maurice, D. M. (1957). The exchange of sodium between the vitreous body and the blood and aqueous humour. *J. Physiol. (London)*, **137**, 110–125

Maurice, D. M. (1959). Protein dynamics in the eye studied with labelled proteins. *Am. J. Ophthalmol.*, **47**, 361–367

Maurice, D. M. (1980). Drug exchanges between the blood and vitreous. In Cunha-Vaz, J. (Ed.), *Blood–Retinal Barriers*, NATO Advanced Study Inst. Series, Vol. 32. Plenum Press, New York, pp. 165–179

Maurice, D. M. (1985). Theory and methodology of vitreous fluorophotometry. *Jpn J. Ophthalmol.*, **29**, 119–130

Maurice, D. M. (1987). The flow of water between the aqueous and vitreous compartments in the rabbit eye. *Am. J. Physiol.*, **21**, F104–F108

Moseley, H. (1981). Mathematical model of diffusion in the vitreous humour of the eye. *Clin. Phys. Physiol. Meas.*, **2**, 175–181

5

Role of Carbonic Anhydrase in Aqueous Humour and Cerebrospinal Fluid Formation

Thomas H. Maren

In 1956 there appeared one of the most remarkable books in the history of physiology. The author ascribed his subject to a 'back water of physiology; academic physiologists over the past century have only rarely interested themselves actively in either AH or CSF' (Davson, 1956). He went on most significantly to 'implicitly confess to what may be an unfounded prejudice, namely that the two fluids are formed by mechanisms that have a great deal in common'.

I took up this problem many years later, and for different reasons, but my work ultimately led to the same conclusion as Hugh Davson's 'unfounded prejudice' of 1956. We gave ourselves the conceit that we were an exclusive club of two: that no one else worked on both fluids. It was certainly a good basis for a deep, lasting friendship, not threatened very much by Hugh's persistent Victorian designation of our innocent and harmless carbonic anhydrase inhibitors as 'poisons'.

THE EYE, THE AQUEOUS HUMOUR AND GLAUCOMA

The history of the physiology and anatomy of the aqueous humour (AH) has been told by Davson (1956, 1980), but it is worth recording for this occasion that not so long ago the aqueous was regarded as a stagnant fluid by the leading ophthalmologist of Great Britain in the definitive text of its time (Duke-Elder, 1932). It was in great part due to Hugh Davson, following the lead of Seidel, that the modern physiology and biochemistry of AH has been elucidated.

I have told elsewhere how the work and ideas of Friedenwald, Wistrand, Kinsey, Becker and Roblin in the 1950s led to the firm base that the enzyme carbonic anhydrase (CA) plays a pivotal role in the secretion of aqueous humour, and that its inhibition reduces flow and pressure in normal animals and patients with glaucoma (Maren, 1984). Importantly also, inhibitors of this enzyme were the first drugs to reduce secretion rate, leaving the facility of outflow untouched.

However, there followed many years of uncertainty about the underlying chemistry, and echoes of this are still found in some current textbooks, which read, 'the mechanism of action of these drugs is obscure'. Nothing could be further from the truth, as I shall show, leaving many details to earlier publications reviewed in Maren (1980, 1984).

All known vertebrates have a HCO_3^- accumulating system in the aqueous humour, based on the reaction $CO_2 + OH^- \rightarrow HCO_3^-$ in the ciliary processes (or ciliary folds in lower animals). The reaction is catalysed by CA but proceeds at a measurable rate both chemically and physiologically without the enzyme. Secretory cells are designated to deliver OH^- at their luminal surface, and CO_2 is delivered from blood or cell metabolism. In many species there is a notable excess of HCO_3^- in the posterior chamber (rabbit; the dogfish, *Mustulus canis*), but the system is at work even when this excess is not present (primate, dog, cat). This general scheme is by no means confined to the ciliary process but extends to the pancreas, corneal endothelium, intestine and other systems throughout nature (Maren, 1980).

Definitive experiments to link ion and fluid transport with this chemical paradigm are those which measure the accession of labelled Na^+, Cl^- and HCO_3^- from plasma to posterior aqueous, in the normal, and following CA inhibition (Zimmerman *et al.*, 1976a,b). They are summarized in Table 5.1. Significant points are:

(1) Of the Na^+ transported, about 40% is linked to HCO_3^-.

(2) Nascent fluid is isotonic with respect to sodium, but HCO_3^- is much higher, and Cl^- lower, than in plasma.

(3) Inhibition of CA eliminates about 70% of HCO_3^- accession and 30% of Na^+ accession. Cl^- accession is unchanged.

(4) The decrease in Na^+ accession is matched by the decrease in aqueous flow.

These data put the scheme outlined above in quantitative perspective. We have also been interested in the relation between the observed rate of HCO_3^- transport (Table 5.1) and the catalytic potential of the cell. To this end we have applied the known chemical rate constants and substrate concentration to the ciliary process cells. When this is done, it yields a rate some thousand times that observed in Table 5.1 (Maren, 1980). A similar calculation for the uncatalysed reaction, on the other hand, gives about the same rate as observed for the inhibited HCO_3^- accession (Table 5.1) *in vivo* (Maren, 1980). Thus, it is clear that the enzyme is present in great excess of physiological needs and validates the finding that over 99% inhibition is necessary to produce a pharmacological effect (Maren, 1963).

I turn now to the pharmacology of inhibition of carbonic anhydrase. This

Table 5.1 Accession of ions to posterior aqueous of dog: effect of carbonic anhydrase inhibition

Ion	(1) Plasma (mM)	(2) Posterior aqueous (mM)	(3) k_{in} (min^{-1})	(4) Accession rate: column (1) × column (3) (mM/min)	(5) Calculated composition new fluid[a] (mM)
Na$^+$					
control	152	153	0.044	6.7	149
CA I			0.031	4.8	149
Cl$^-$					
control	117	131	0.028	3.3	73
CA I			0.027	3.2	100
HCO$_3^-$					
control			2.4	2.4	53
	22	25			
CA I			0.8	0.8	25

[a] Column (4) × $\dfrac{\text{volume posterior chamber (0.3 ml)}}{\text{flow}}$.

has been reviewed many times (Maren, 1967, 1987) and will be summarized briefly. All drugs of this type are unsubstituted sulphonamides, $R\text{-}SO_2NH_2$, where R is aromatic or heteroaromatic. They are reversible, non-competitive (with CO_2) inhibitors, with K_I ranging from 10^{-5} to 10^{-10} M. Figure 5.1 shows the structures of the four compounds that have come into use as parenteral treatment for glaucoma. The best-known is acetazolamide, but it is likely that methazolamide (Maren *et al.*, 1977) is the better drug, in terms of balance between toxicity and efficacy. The relation between pharmacological properties and intraocular effects of nine sulphonamides was worked out by Wistrand *et al.* (1961) many years ago, in a neglected series of papers, well worth reading.

In the last 10 years attention has turned to new sulphonamides which can be given topically and thus avoid systemic effects of the 'classic' parenteral drugs of Figure 5.1. This figure also shows derivatives of these compounds which held promise as topical drugs. The development of this new field is told in Maren (1987) and Maren *et al.* (1990) and will be outlined here.

The desirable characteristics for new compounds to be topically active appeared to be (1) high activity against the enzyme; (2) moderate water-solubility; (3) moderate lipid-solubility; (4) good transcorneal permeability *in vivo* as well as *in vitro*. For one reason or another, none of the four compounds of Figure 5.1 fit these criteria; it was commonly held that a drug of this type was an impossibility, since none had been developed from 1955 to 1980.

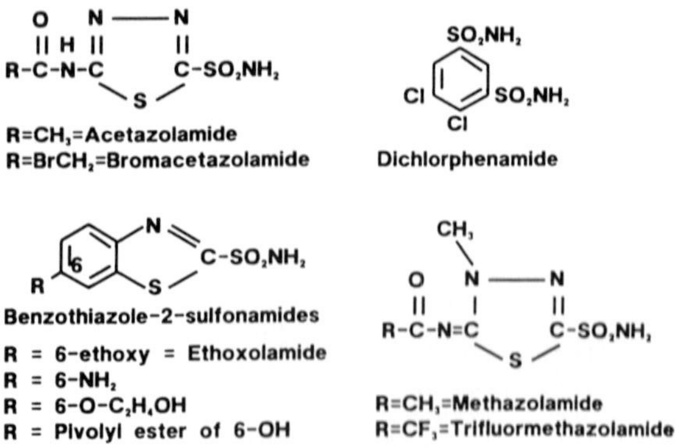

Figure 5.1 Structures of 'classical' carbonic anhydrase inhibitors and derivatives developed in the search for a topically active compound

However, new work showed that such activity was possible (Maren *et al.*, 1983), and a long series of studies at the University of Florida and at the Merck Sharp & Dohme Research Laboratories culminated in new sulphonamides of the structures shown in Figure 5.1, all synthesized at Merck (Sugrue *et al.*, 1960, 1990; Baldwin *et al.*, 1989; Maren *et al.*, 1990).

Table 5.2 shows the properties of MK-927, a prototype drug of the ampholytic series shown in Figure 5.2, which contain a basic $-NH_2$ group and an acidic $-SO_2NH_2$ group. Note high activity against CA II, the secretory cytoplasmic enzyme, and CA IV, the secretory membrane bound enzyme, but virtual inactivity against CA I, the second isozyme of red cells and chief CA of the intestine. Also shown are the water- and lipid-solubility at different pHs, corresponding to the two pK_as of the compound.

Figure 5.3 shows the pressure-lowering effect of MK-927 in the pigmented rabbit and the curious still unexplained fact that the effect is the greater the lower the pH. This approaches or equals that of systemic acetazolamide or methazolamide. Table 5.3 shows the distribution of MK-927 in the albino rabbit following a single drop of 2% drug. Note the concentration of drug in the aqueous humour, uptake into the cornea and presence in the ciliary process in μM concentration. Since the K_I against

Table 5.2 Physicochemical properties of MK-927

K_I *vs CA* (nM)		*pK*$_a$	*% ionized*	*Solubility* (mM)	CHCl₃/ *Buffer ratio*	k_{in} *cornea* $\times 10^3$ h^{-1}	
II, IV	*I*					*in vitro*	*in vivo*
0.5–2	~10 000	5.8 pH 5	86	60	0.3	0.4	4
		8.3 pH 7	6	13	2.0	3.0	5

L-650,719

L-662,583

MK-927

MK-417

MK-507

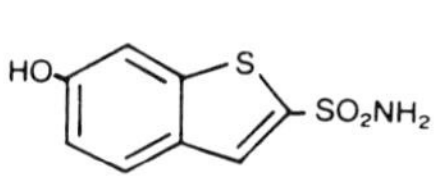

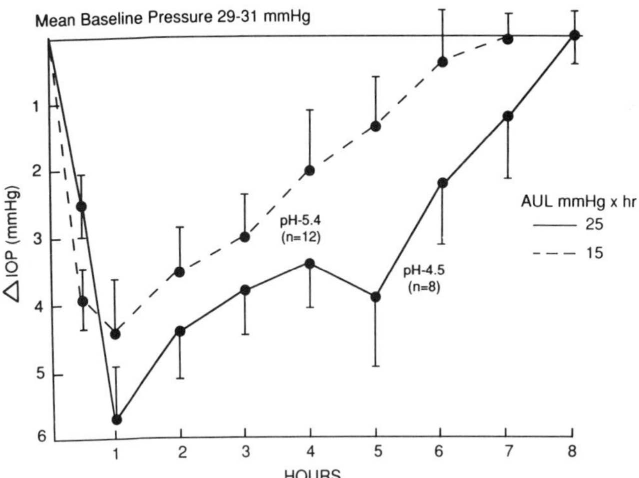

Figure 5.2 Sulphonamides developed (1986–1989) in the search for topical treatment of glaucoma (Merck)

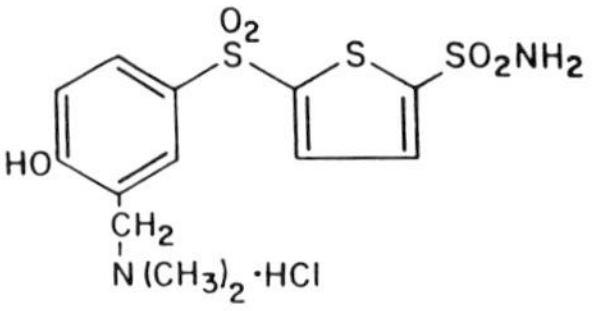

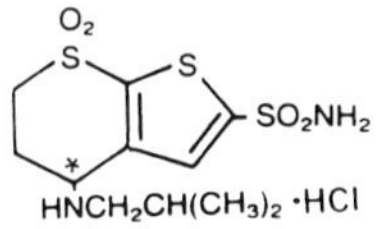

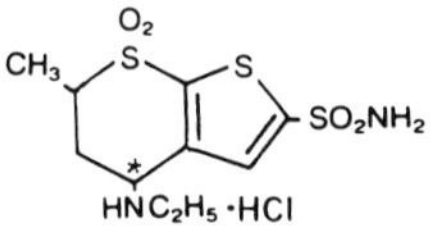

enzyme is in the nM range (Table 5.2), it is evident that inhibition may be achieved by this route.

At the present time MK-507, which appears more active than MK-927 (Sugue *et al.*, 1990), is in widespread clinical trial, and, it is hoped, will be marketed within the next few years.

Figure 5.3 The lowering of intraocular pressure in the pigmented rabbit by MK-927. This group of animals, as noted, have spontaneous ocular hypertension. Drug given in 2% solution at the pH noted, in 0.5% hydroxyethylcellulose. Δ IOP is difference in pressure between treated and untreated eye. AUL is area under the line

Table 5.3 Distribution and decay of MK-927 in albino rabbit following one drop of 2% solution (pH 4.5) in 0.5% HEC. Concentration in μM or $\mu mol/kg \pm SE$

Hours	Aqueous		Cornea	Anterior uvea	Ciliary process
	anterior	posterior			
0.5	20 ± 2^a	—	160 ± 10	12 ± 2	—
1	32 ± 3	4 ± 1	108 ± 10	32 ± 8	15 ± 3
3	19 ± 2	3 ± 1	56 ± 14	17 ± 2	11 ± 1
9	3 ± 0.2	2 ± 0.3	8 ± 1	10 ± 2	8 ± 1
24	<0.5	2 ± 1	<1	4 ± 1	4
$t_{1/2}$ (h)[b]	3	8, ?	3	2, 10	8

[a] Concentration at 5 min = 3; at 10 min = 8; at 15 min = 21.
[b] Calculated for periods after 1 h.
n = 6–10.

CEREBROSPINAL FLUID (CSF)

History of progress on the physiology of CSF is quite different from that on the physiology of AH. In 1956 Hugh Davson thought of himself rather as an interloper in this field, after some 20 years' work on the aqueous. He supposed, rather tentatively, that the two fluids were formed by similar mechanisms, but there was still a school of thought that maintained that CSF was formed by nervous tissue rather than choroid plexus, and he was impressed by the Cl^- excess in CSF in all species—not the case for AH.

Not surprisingly, Hugh Davson soon achieved mastery of this field as well, building on the ideas of Lewis Weed. His ideas and experiments culminated in the great yellow book (Davson, 1967) entirely devoted to the CSF.

Soon after acetazolamide was developed in the early 1950s, it was found that systemic administration reduced CSF flow (Tschirgi *et al.*, 1954; Kister, 1956) and CA was found in the choroid plexus (Birzis *et al.*, 1958). Davson and Luck (1957) then showed a reduction in Na^+ accession by acetazolamide. This became part of the central dogma for both AH and CSF, that Na^+ and flow moved together, and nascent CSF (as well as AH) had the same Na^+ concentration as plasma.

Still, the connection with carbonic anhydrase was puzzling. There was no HCO_3^- excess in CSF, as for AH of some species; the rule was a slight HCO_3^- deficit and Cl^- excess. In the dogfish, *Squalus acanthias*, carbonic anhydrase inhibition caused a lowering of AH HCO_3^- (as in the rabbit) but a very pronounced (fourfold) rise in CSF HCO_3^- (Birzis *et al.*, 1958). It appeared then as if the chemistry underlying the two fluids was very different. When I reviewed this matter in 1967, I came to the rather lame conclusion that we were dealing with an acid secretion somehow mediating

Cl^- transport (Maren, 1967). To try to allay this unsatisfactory situation, we began new experiments in the cat and the dogfish.

Ventriculocisternal perfusion in the cat was accompanied by intravenous injection of $H^{14}CO_3^-$ and $^{36}Cl^-$. Both ions appeared rapidly in the perfusate, and acetazolamide, quite unexpectedly, reduced the accession of both (Maren and Broder, 1970). This was the first experiment linking HCO_3^- movement in CSF to carbonic anhydrase. The next and crucial work was done in the dogfish and is shown in Figure 5.4. If I may be permitted to have favourites among experiments, this is surely one. On the left is shown the remarkable effect of CO_2 in elevating CSF HCO_3^-; on the right is the same type experiment but following inhibition of carbonic anhydrase. The rate of CSF HCO_3^- formation is halved. The only explanation is that the $CO_2 + OH^- \rightarrow HCO_3^-$ system is at work, just as for AH and pancreas. The early experiments in which CA inhibition caused the puzzling rise in CSF concentration (Maren, 1962) could now be explained by the marked respiratory acidosis in the fish, secondary to inhibition of red cell CA. High P_{CO_2} (4 times normal) drove HCO_3^-

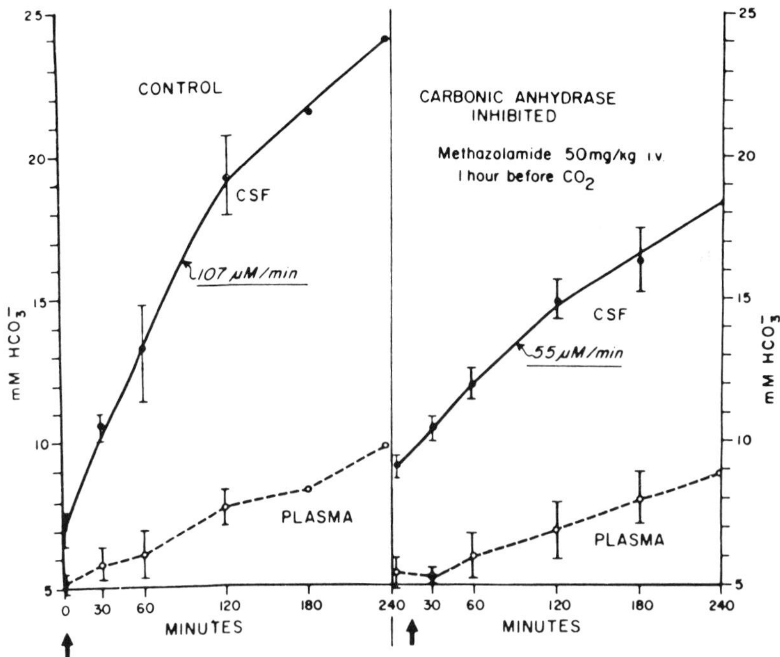

Figure 5.4 Effect of raising blood P_{CO_2} from 4 to 16 mmHg in the dogfish, *Squalus acanthias*. Mean $\pm$ SE of all experiments. Controls, $n = 5$; inhibited, $n = 4$. For points past 180 min, $n = 2.5\%$ CO_2 in gill perfusate at large arrow. Numbers alongside small arrow give mean rates

formation even when choroid plexus enzyme was inhibited, as shown on the right of Figure 5.4.

With the vital clue of Figure 5.4, everything fell into line. The CO_2 effect on CSF HCO_3^- so prominent in fish (Figure 5.4) may also be observed in the mammal (Figure 5.5); note the lack of HCO_3^- formation in blood or skeletal muscle. Figure 5.5 also shows that CO_2 elicits HCO_3^- formation in brain; presumably this is to serve pH homoeostasis. The same prominent effect is seen in cardiac muscle (Lai *et al.*, 1973).

Accession rates to CSF of the ions Na^+, Cl^- and HCO_3^- were studied by means of their isotopes, similar to what had been done for the AH. In both dogfish (Maren, 1972) and cat (Vogh and Maren, 1975) sodium was transferred isotonically, but nascent HCO_3^- was high and Cl^- low, just as for AH. Table 5.4 shows data for the cat, including the effect of CA inhibition, which lowered the ionic transfers as well as the nascent HCO_3^- concentration. Comparison with Table 5.1 shows the similarity to AH data, with the important (and as yet unexplained) exception that, for CSF, inhibition lowers Cl^- transfer, as observed earlier under quite different conditions (Maren and Broder, 1970).

Thus, the CSF secretory mechanism is, after all, quite analogous to that of AH, pancreas and other organs that use HCO_3^- formation, linked to

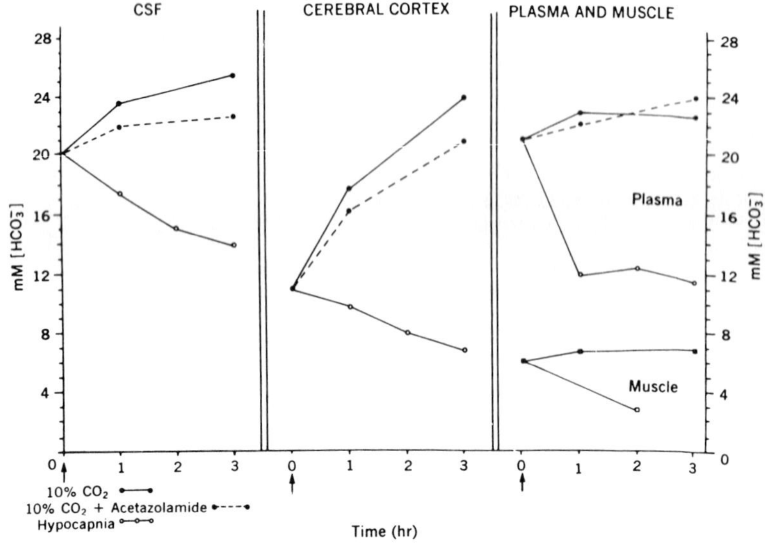

Figure 5.5 Effects of hypercapnia (10% CO_2, arterial P_{CO_2} = 85 mmHg), shown in solid points, and of hypocapnia (mechanical ventilation adjusted to arterial P_{CO_2} = 15 mmHg), shown in open points, on HCO_3^- concentration of CSF, cerebral cortex, plasma and muscle of dogs. Abscissae show time from zero of altered P_{CO_2}. Dotted lines show experiments in which dogs were pretreated with intravenous acetazolamide (100 mg/kg at 0 and 2 h) to inhibit completely carbonic anhydrase. Based on data from Arieff *et al.* (1976)

Table 5.4   Accession of ions to CSF of cat: effect of carbonic anhydrase inhibition

Ion	(1) *Plasma* H_2O (mM)	(2) *CSF* (mM)	(3) k_{in} (min^{-1})	(4) *Accession rate: column (1) × column (3)* (mM/min)	(5) *Calculated composition new CSF^a* (mM)
Na^+					
control			0.016	2.4	166
	147	158			
CA I			0.007	1.1	150
Cl^-					
control			0.013	1.5	104
	115	134			
CA I			0.006	0.7	101
HCO_3^-					
control			0.044	0.9	62
	20	22			
CA I			0.014	0.3	49

^{a} Column (4) × $\dfrac{\text{volume CSF (1.4 ml)}}{\text{flow}}$.

sodium, as a means of moving fluid. It is even the case that the capacity to form HCO_3^-, factored by cell volume, is very similar among the various organs (Maren, 1980).

As the foregoing data show, inhibition of CA reduces Na^+, HCO_3^- and flow, but not completely. The question has continuously arisen as to the nature of the remainder. An obvious candidate is the uncatalysed reaction of CO_2 to yield HCO_3^-. Although this can be calculated readily, the extension to *in vivo* rates is, while interesting, treacherous because of uncertainties in knowing cell volumes and substrate gradients. We therefore sought an experimental value for the uncatalysed rates.

Ventriculocisternal perfusion was carried out in the rat, to measure CSF flow (Vogh *et al.*, 1985, 1987). To the perfusion fluid was added Al^{3+} or Ga^{3+} or acids: the pH was lowered to 3–5. In separate experiments acetazolamide was given systematically. Table 5.5 shows the results. Acid perfusions lowered flow 30%; acetazolamide, 43%. The two together lowered flow 64%; adding the effects yielded 73%. All effects were reversible. Our conclusion is that the acid infusions abolish the uncatalysed rate but not the catalysed. The basis for this is direct dependency of the uncatalysed rate on OH^- and CO_2: if cell pH were lowered two pH units, the uncatalysed rate would be 1% of normal. The catalysed rate would also be lowered, but since there is some thousandfold excess of enzyme, this would not be apparent in the *in vivo* rate. We are greatly indebted once again to Hugh Davson and his new young colleague, Berislav Zloković, for

Table 5.5 Effect of acids and CA I on CSF flow in rat

Acid/CA I	n	CSF flow (μl/min $\pm$ SE)
Control	16	3.0 ± 0.08
$AlCl_3$	19	2.1 ± 0.09
$GaCl_3$	4	2.2 ± 0.2
HCl, H_3PO_4, HAc	15	2.0 ± 0.2
Acetazolamide i.v.	9	1.7 ± 0.1
$AlCl_3$ + acetazolamide	4	1.1 ± 0.1
Summation: acetazolamide + acid		0.8

pointing out a serious error in our first paper (Vogh *et al.*, 1985); their experiments (Zloković *et al.*, 1987) agree with our corrected version (Vogh *et al.*, 1987), as shown in Table 5.5.

We are left with what at first seems paradoxical but may be an important clue to one of the secrets of secretion. If our analysis of Table 5.6 is correct, 73% of flow is dependent on HCO_3^- passage, linked, of course, to sodium. But Table 5.4 shows that only 37% of the sodium is matched by HCO_3^-, the remainder being matched by Cl^-. Thus, it appears that the HCO_3^- moiety carries more of its share of water than does chloride—i.e. there is a special relation between HCO_3^- and flow. We have seen this relation also in AH secretion. The most obvious example is in pancreatic secretion, which is turned on when secretin elicits a shift from Cl^- to HCO_3^- output.

We then are led to ask: what links HCO_3^- to Na^+? If the ions were free, discrimination between Cl^- and HCO_3^- with respect to flow would seem unlikely. We inquired whether ion pairs between Na^+ and HCO_3^- could account for the physiological observations (Conroy and Maren, 1989; Table 5.6). In aqueous solution the association of the ions is 8%, perhaps negligible in ordinary considerations of acid–base balance. However, at lower dielectric constant (ϵ) the association increases markedly, to 98% at $\epsilon = 15$, the approximate value in the vicinity of cell membranes (Monoi, 1982). Under this circumstance the flux of $[^{22}Na^+]$–HCO_3^- is increased some sixtyfold over the flux in water (Table 5.6), suggesting the possibility that this may occur *in vivo*. In essence the dehydrated ion pair, $NaHCO_3^\circ$, moves across the membrane of low dielectric constant. Upon movement to a region of higher dielectric constant beyond the membrane, the ions regain their original waters of hydration, which inevitably leads to a net movement of water. Thus, the link between Na^+ and HCO_3^- is chemical, rather than physiological, and the carbonic anhydrase system is so important because it furnishes the ion that can link with Na^+ and move fluid.

In conclusion, I have tried to show the role of HCO_3^- formation in the secretion of AH and CSF and the link with sodium formation and fluid movement. The physiology of AH and that of CSF are remarkably similar, as foretold by Hugh Davson. The development of specific sulphonamide

Table 5.6 The effect of lowered dielectric constant and ionization on the formation of ion pair and flux of $NaHCO_3$ across a lipophilic membrane

Medium	Dielectric constant	$^{22}NaHCO_3$ flux $\times$ 10^{10} mol cm^{-2} s^{-1}	K_A, M^{-1} $NaHCO_3^\circ$	% $NaHCO_3$ associated[c]
Water	78	0.07	0.6[a]	8
Dioxane 50%	36	2.6	2.6[b]	28
Dioxane 75%	15	4.3	340[b]	98
Dioxane 90%	6	6.8	3.6×10^{5}[b]	>99.9

[a] From conductivity measurements.
[b] Calculated from Debye–Hückel theory and the Bjerrum equation.
[c] $NaHCO_3^\circ$/total $NaHCO_3 \times 100$. From $NaHCO_3^\circ = K_A \cdot (Na^+) \cdot (HCO_3^-$, where $Na^+ = 0.15$ M and total $HCO_3^- = 0.015$ M.

inhibitors of CA has made possible much of our understanding of this field, and the pharmacology of new drugs of this class gives further promise in the treatment of glaucoma.

ACKNOWLEDGEMENTS

This work was largely supported over 25 years by grants from the National Eye Institute of the National Institutes of Health.

REFERENCES

Arieff, A. I., Kernian, A., Massry, S. G. and Delima, J. (1976). Intracellular pH of brain: alterations in acute respiratory acidosis and alkalosis. *Am. J. Physiol.*, **230**, 804

Baldwin, J. J., Ponticello, G. S., Anderson, P. S., Christy, M. E. *et al.* (1989). Thienothiopyran-2-sulfonamides: Novel topically active carbonic anhydrase inhibitors for the treatment of glaucoma. *J. Med. Chem.*, **32**, 2510–2513

Birzis, L., Carter, C. H. and Maren, T. H. (1958). Effect of acetazolamide on CSF pressure and electrolytes in hydrocephalus. *Neurology*, **8**, 522–528

Conroy, C. W. and Maren, T. H. (1989). The permeability of hydrophobic membranes to ^{22}Na salts and $^{14}CO_2$ in low dielectric media. *Biophys. Chem.*, **34**, 177–184

Davson, H. (1956). *Physiology of the Ocular and Cerebrospinal Fluids*. Little, Brown, Boston

Davson, H. (1967). *Physiology of the Cerebrospinal Fluid*. Churchill, London

Davson, H. (1980). *Physiology of the Eye*, 4th edn. Academic Press, New York

Davson, H. and Luck, C. P. (1957). The effect of acetazolamide on the chemical composition of the aqueous humor and cerebrospinal fluid of some mammalian species and on the rate of turnover of ^{24}Na in these fluids. *J. Physiol. (London)*, **137**, 279–293

Duke-Elder, W. S. (1932). *A Textbook of Ophthalmology*, Vol. 1. Kimpton, London

Kister, S. (1956). VI. The effect of acetazolamide on cerebrospinal fluid flow. *J. Pharmacol. Exp. Ther.*, **117**, 402–406

Lai, Y. L., Attebery, B. A. and Brown, E. B., Jr. (1973). Intracellular adjustments of skeletal muscle, heart, and brain to prolonged hypercapnia. *Resp. Physiol.*, **19**, 115–122

Maren, T. H. (1962). Ionic composition of cerebrospinal fluid and aqueous humor of the dogfish, *Squalus acanthias*. II. Carbonic anhydrase activity and inhibition. *Comp. Biochem. Physiol.*, **5**, 201–215

Maren, T. H. (1963). The relation between enzyme inhibition and physiological response in the carbonic anhydrase system. *J. Pharmacol. Exp. Ther.*, **139**, 140–153

Maren, T. H. (1967). Carbonic anhydrase: Chemistry, physiology, and inhibition. *Physiol. Rev.*, **47**, 595–781

Maren, T. H. (with the technical assistance of Kent, B. B., Welliver, R. C. and Woodworth, R. B.) (1972). Bicarbonate formation in cerebrospinal fluid: role in sodium transport and pH regulation. *Am. J. Physiol.*, **222**, 885–899

Maren, T. H. (1980). The kinetics of HCO_3^- synthesis related to fluid secretion, pH control, and CO_2 elimination. *Ann. Rev. Physiol.*, **50**, 695–717

Maren, T. H. (1984). The development of ideas concerning the role of carbonic anhydrase in the secretion of aqueous humor: Relation to the treatment of glaucoma. In Drance, S. M. and Neufeld, A. H. (Eds), *Glaucoma: Applied Pharmacology in Medical Treatment.* Grune and Stratton, Orlando, Florida, pp. 325–355

Maren, T. H. (1987). Carbonic anhydrase: General perspectives and advances in glaucoma research. *Drug Devl Res.*, **10**, 255–276

Maren, T. H., Bar-Ilan, A., Conroy, C. W. and Brechue, W. F. (1990). Chemical and pharmacological properties of MK-927, a sulfonamide carbonic anhydrase inhibitor that lowers intraocular pressure by the topical route. *Exp. Eye Res.*, **50**, 27–36

Maren, T. H. and Broder, L. E. (1970). The role of carbonic anhydrase in anion secretion into cerebrospinal fluid. *J. Pharmacol. Exp. Ther.*, **172**, 197–202

Maren, T. H., Haywood, J. R., Chapman, S. K. and Zimmerman, T. J. (1977). The pharmacology of methazolamide in relation to the treatment of glaucoma. *Invest. Ophthalmol. Vis. Sci.*, **16**, 730–742

Maren, T. H., Jankowska, L., Sanyal, G. and Edelhauser, H. F. (1983). The transcorneal permeability of sulfonamide carbonic anhydrase inhibitors and their effect on aqueous humor secretion. *Exp. Eye Res.*, **36**, 457–480

Monoi, H. (1982). Possible existence of ion pairs at the mouths of ion channels. *Biochim. Biophys. Acta*, **693**, 159–164

Sugrue, M. F., Gautheron, P., Mallorga, P., Nolan, T. E., Graham, S. L., Schwam, H., Shepard, K. L. and Smith, R. L. (1990). L-662,583 is a topically effective ocular hypotensive carbonic anhydrase inhibitor in experimental animals. *Br. J. Pharmacol.*, **99**, 59–64

Sugrue, M. F., Mallorga, P., Schwam, H., Baldwin, J. J. and Ponticello, G. S. (1960). A comparison of L-671,152 and MK-927, two topically effective ocular hypotensive carbonic anhydrase inhibitors in experimental animals. *Curr. Eye Res.* (in press)

Tschirgi, R. D., Frost, R. W. and Taylor, J. L. (1954). Inhibition of cerebrospinal fluid formation by a carbonic anhydrase inhibitor (Diamox). *Proc. Soc. Exp. Biol. Med.*, **87**, 373–376

Vogh, B. P., Godman, D. R. and Maren, T. H. (1985). Aluminum and gallium arrest formation of cerebrospinal fluid by the mechanism of OH^- depletion. *J. Pharmacol. Exp. Ther.*, **233**, 715–721

Vogh, B. P., Godman, D. R. and Maren, T. H. (1987). The effect of $AlCl_3$ and other acids on cerebrospinal fluid production: A correction. *J. Pharmacol. Exp. Ther.*, **243**, 35–39

Vogh, B. P. and Maren, T. H. (1975). Sodium, chloride, and bicarbonate movement from plasma to cerebrospinal fluid in cats. *Am. J. Physiol.*, **228**, 673–683

Wistrand, P. J., Rawls, J. A., Jr. and Maren, T. H. (1961). Sulfonamide carbonic anhydrase inhibitors and intra-ocular pressure in rabbits. *Acta Pharmacol. Toxicol.*, **17**, 337–355

Zimmerman, T. J., Garg, L. C., Vogh, B. P. and Maren, T. H. (1976a). The effect of acetazolamide on the movements of anions into the posterior chamber of the dog eye. *J. Pharmacol. Exp. Ther.*, **196**, 510–516

Zimmerman, T. J., Garg, L. C., Vogh, B. P. and Maren, T. H. (1976b). The effect of acetazolamide on the movement of sodium into the posterior chamber of the dog eye. *J. Pharmacol. Exp. Ther.*, **199**, 510–517

Zloković, B. V., Davson, H., Preston, J. E. and Segal, M. B. (1987). The effects of aluminum chloride on the rate of secretion of cerebrospinal fluid. *Exp. Neurol.*, **98**, 436–452

6
Pharmacological Manipulation of Cerebrospinal Fluid Secretion

Michael Pollay

INTRODUCTION

Our understanding of the site and volume of cerebrospinal fluid formation is due in large measure to the development of the ventriculocisternal perfusion system. This experimental technique allowed the accurate measurement of both the volume of fluid entering and leaving the ventricular system and the dilution of a non-diffusible marker in transit through this experimental system (Pappenheimer *et al.*, 1961). In the past 30 years, the manipulation of the rate of CSF formation with various pharmacological agents has allowed some understanding of the mechanisms involved in the secretory process. The discussion that follows represents the published and unpublished experimental results describing the effect of cardiac glycosides, steroids and benzodiazepine ligands on the secretion of CSF within the ventricular system.

CARDIAC GLYCOSIDES AND CSF SECRETION

In 1963 Davson and Pollay (1963), utilizing a ventriculocisternal perfusion system in the rabbit, demonstrated that cerebrospinal fluid (CSF) secretion was inhibited some 21.6% when 0.01 mg/100 ml of ouabain was added to the fluid perfusing the ventricular system. This observation supported the notion that, as in other secretory systems, this effect was due primarily to the inhibition of Na^+/K^+-activated ATPase and the active transport of sodium across the secretory membrane (Davson and Segal, 1970; Pollay *et al.*, 1985). In the latter paper Pollay *et al.* examined the relationship between the concentration of ouabain in ventricular CSF and the degree of inhibition of both CSF formation and choroid plexus ATPase activity. In both dog and rabbit, Na^+/K^+-activated ATPase represents, respectively, between 20% and 27% of the total ATPase activity. This component in plexus tissue is totally inhibited by a concentration of 10^{-4} M ouabain. At this same concentration of glycoside in CSF, the rate of CSF formation fell by some 70–80%. The remaining 20–30% of intraventricular CSF formation is unaffected, which suggests either that there are two different mechanisms responsible for intraventricular fluid formation or that a significant portion of CSF formation is from an extrachoroidal source not utilizing sodium–potassium-activated ATPase in the secretory process (Pollay *et al.*, 1985). The latter option is compatible with studies demonstrating intraventricular CSF formation in the absence of choroid plexus tissue (Pollay and Curl, 1967).

EFFECT OF DEXAMETHASONE ON CSF SECRETION AND ATPase

It is agreed that steroids are useful in treating cerebral oedema by decreasing the volume of excess brain water found in this condition, although it is unclear that the lowering of intracranial pressure is due to this process alone, since the volume of the CSF compartment may also be constricted (Long *et al.*, 1966; Reulen and Kreysch, 1973). This effect on CSF volume has been unevenly supported by the notion that ventricular fluid formation is significantly inhibited by steroidal agents (Amano, 1969; Garcia-Bengochea, 1965; Lindvall-Axelsson *et al.*, 1989; Martins *et al.*, 1977; Sato, 1967; Sato *et al.*, 1973; Vela *et al.*, 1979; Weiss and Nulsen, 1970), presumably owing to suppression of Na^+/K^+-activated ATPase activity in choroid plexus (Mayman, 1972; Lindvall-Axelsson *et al.*, 1989). Because of the conflicting experimental data available, we experimentally determined the effect of systemic and intraventricular dexamethasone on the production of CSF *in vivo* and plexus ATPase *in vitro*. The influence of this steroid on CSF formation was then related to changes in choroid plexus ATPase activity at the same concentration levels.

Bilateral ventriculocisternal perfusion systems were established in 25 adult, anaesthetized, paralysed and artificially ventilated, rabbits. The ventricular system was perfused at a rate of 105 µl/min with a gassed bicarbonate-buffered artificial CSF solution containing a non-diffusible marker ([³H]-dextran). Outflow from the fourth ventricle was collected at 10 min intervals over a 5 h period. The rate of CSF formation (V_f) was calculated from the fluid volumes entering and leaving the ventricular system and the dilution of the radioactive marker (Pollay and Curl, 1967). Each animal served as its own control by comparing V_f during the initial perfusion period (150 min) with that following the addition of dexamethasone to the perfusate (10^{-3} M and 10^{-5} M) or by intravenous infusion (1.0 or 10.0 mg/kg body weight). The measured values of V_f were corrected for deterioration in the perfused specimen that has been reported to occur in ventricular perfusion systems in the rabbit (Pollay *et al.*, 1985). In five animals, the mean rate for V_f fell from 8.48 ± 0.55 μl/min to 7.43 ± 0.29 μl/min over a 5 h period, which is about a 2.0% reduction in V_f per hour. In these same animals, total ATPase activity (based on dry weight) was found to be about 15 mol kg^{-1}h^{-1}, which is similar to that previously reported in non-perfused rabbits (Pollay *et al.*, 1985). The total ATPase activity was measured in pooled lateral and fourth ventricular choroid plexus, using the colorimetric method of Lanzetta *et al.* (1979) and expressed as nmol of inorganic phosphate (Pi) released (by hydrolysis of ATP) per 30 μg of dry weight of choroid plexus tissue per min. The concentration of dexamethasone added to the prepared plexus tissue varied between 10^{-7} and 10^{-3} M. In order to evaluate the effect of this steroid on the two fractions of ATPase, 10^{-3} M ouabain was used to inhibit the Na$^+$/K$^+$-activated ATPase fraction. The Ca^{2+}/Mg^{2+}-activated ATPase was computed from the difference between total ATPase activity and that requiring sodium and potassium activation.

The results of these experiments are shown in Table 6.1.

It can be seen that after correcting for a decrease in CSF formation (V_f) due to deterioration of the experimental perfusion system the inhibitory effect of intravenous dexamethasone on V_f is only significant at a dose of 1.0 and 10.0 mg/kg when intraventricular fluid formation fell by 40% and 34%, respectively. When added to the perfusion fluid, this steroid inhibited V_f by 20% and 26% at concentration levels of 10^{-5} M and 10^{-3} M, although only the higher dose was found to be statistically significant as compared with control values ($p < 0.05$). In the *in vitro* experiments, the control plexus tissue (no dexamethasone added) revealed a total ATPase activity of 9.5 nmol Pi/min in 30 μg of dry tissue, while the Ca^{2+}/Mg^{2+}- and Na$^+$/K$^+$-activated fractions were 7.1 and 2.9 nmol Pi/min, respectively. It

Table 6.1 Effect of dexamethasone on CSF formation

Dosage[a]	Pre-treatment V_f^b	Post-treatment V_f^b	% Decrease	p^c
0.1 mg/kg(4)[d]	8.10 ± 0.65	6.82 ± 1.08	17%	>0.05
1.0 mg/kg(4)[d]	7.10 ± 0.79	4.28 ± 0.66	40%	<0.05
10.0 mg/kg(4)[d]	7.47 ± 1.10	4.92 ± 0.76	35%	<0.05
10^{-5}M(4)[e]	9.45 ± 1.41	7.58 ± 0.73	20%	>0.05
10^{-3}M(4)[e]	9.78 ± 0.57	7.33 ± 0.67	26%	<0.05

[a] Dexamethasone concentration (number of animals).
[b] Mean ± standard error of the mean.
[c] Student's *t*, significant $p \leq 0.05$.
[d] Intravenous route of administration.
[e] Intraventricular route of administration.

was found that total choroid plexus ATPase was inhibited by some 40% by 10^{-3} M dexamethasone (Figure 6.1).

The shape of this curve and the percentage decrease in the Ca^{2+}/Mg^{2+} fraction (10^{-3} M ouabain in the incubation media) were similar to those determined for total ATPase. A purified sodium–potassium-activated fraction was obtained from brain tissue (Grade III) and demonstrated a maximum inhibition of about 30% when the concentration of dexamethasone was 10^{-3} M (Figure 6.2).

A summary of the results from previous studies on the effect of steroids on CSF formation is presented in Table 6.2.

The earlier studies on the effect of steroids on CSF formation, utilizing dogs and cats, almost uniformly demonstrate a 50% decrease in CSF formation at doses that were a fraction of what we recently found in the experiments reported in this paper (Amano, 1969; Garcia-Bengochea, 1965; Sato, 1967; Sato *et al.*, 1973; Weiss and Neilsen, 1970). As noted by Martins *et al.* (1977) and our experience in the rabbit, there is a gradual and significant (2%/h) decrease in CSF formation and therefore an overestimate of the inhibition of fluid formation in the presence of a putative inhibitor if uncorrected for this deterioration. This may, in part, explain the quantitative differences between our studies in the rabbit and those of Martins *et al.* (1977) and Vela *et al.* (1979) in monkey and dog, respectively, and those studies in which such corrections for nomal decay in fluid formation were not considered (Amano, 1969; Davson and Segal, 1970; Garcia-Bengochea, 1965; Pappenheimer *et al.*, 1961; Pollay and Curl, 1967; Quinton *et al.*, 1973). The studies in both dog and monkey

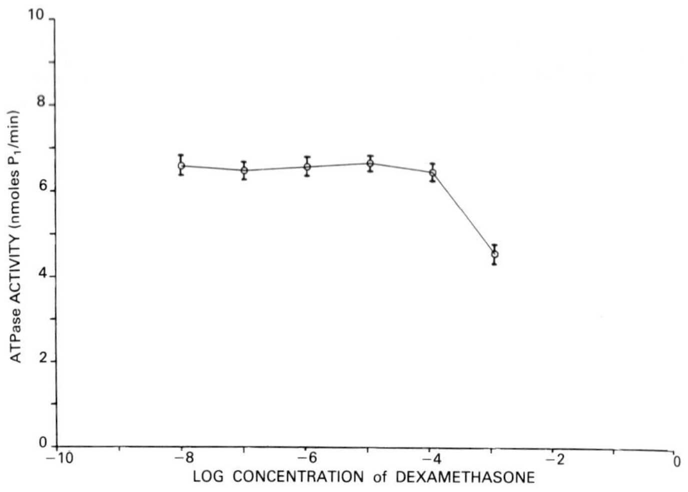

Figure 6.1   The effect of dexamethasone on total choroid plexus ATPase. Values represent mean ± SEM

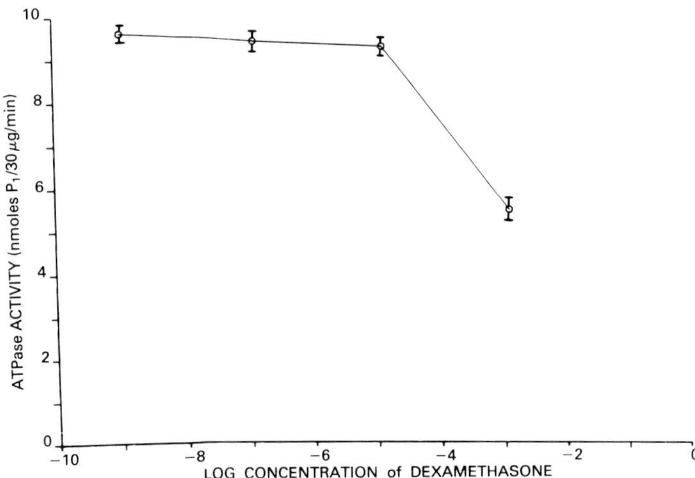

Figure 6.2 The effect of dexamethasone on sodium–potassium-activated ATPase purified from brain. Values represent mean ± SEM

seem to indicate the lack of effect of intravenous dexamethasone on CSF secretion in a dosage below 0.5 mg/kg (Long *et al.*, 1966; Quinton *et al.*, 1973). This essentially is in agreement with the observation in rabbit, where the effective intravenous dose was found to be in excess of

Table 6.2 Summary of studies on the effect of steroids and CSF formation

Experimental model	Dosage	Effect on V_f[a]	Reference
Cat: drop collection	20 mg kg^{-1} d^{-1} i.m. × 14 d	Decrease non-castrated	Garcia-Bengochea (1965)
Dog: VC perfusion	0.15 mg/kg i.v.[c]	50% decrease	Sato (1967); Sato *et al.* (1973)
Dog: brain oedema VC perfusion	0.15 mg/kg i.v.[c]	50% decrease	Amano (1969)
Dog: hydrocephalus gravimetric	0.25 mg/kg i.v.[c] 1.75 mg/kg i.v.[d]	c. 50% decrease c. 50% decrease	Weiss and Nulsen (1970)
Monkey: VC perfusion	0.15 and 0.40 mg/kg i.v.[c]	no significant change	Martins *et al.* (1977)
Dog: VC perfusion	0.40 mg/kg i.v.[c]	no significant change	Vela *et al.* (1979)
Rabbit: VC perfusion	oral 2 mg kg^{-1} d^{-1} × 5 d + 0.5 mg/kg s.q. 2 × d × 2 d[e]	43% decrease	Lindvall-Axelsson *et al.* (1989)
Rabbit: VC perfusion	1–10 mg/kg i.v.[c] 10^{-3} M in CSF	30–40% decrease 26% decrease	PS[f]

[a] V_f represents rate of CSF formation.
[b] Intramuscular cortisone acetate.
[c] Intravenous dexamethasone.
[d] Intravenous methyl prednisolone.
[e] Oral and subcutaneous betamethasone.
[f] Dexamethasone, present study.

1.0 mg/kg. Kinetic studies of the distribution of this steroid between plasma and brain suggest that to reach effective inhibitory concentrations in CSF would require doses in the range of 6 mg/kg (Cham *et al.*, 1980). Betamethasone appears also effective in relatively large oral doses (2.0 mg/kg) supplemented by subcutaneous injection of this steroid (Lindvall-Axelsson *et al.*, 1989). In addition, both agents adversely affect ATPase activity in these dose ranges.

The earliest information available as to the effect of dexamethasone on Na^+/K^+-activated ATPase activity was briefly presented in abstracts by Mayman (1972) and Mayman and Tijerina (1978). In these studies there was a 80–90% inhibition of the enzyme in man, cat and rabbit at concentrations of 10^{-6}–10^{-4} M. We have shown in rabbit choroid plexus an *in vitro* inhibition of some 40% of total ATPase activity (Ca^{2+}/Mg^{2+}- and Na^+/K^+-activated fractions) when the media concentration was 10^{-3} M. At that same concentration level the purified sodium–potassium-activated ATPase activity decreased by about 30%. The observation that the activity of both fractions of choroid plexus ATPase are inhibited by this steroid suggests a non-specific membrane effect, which is in contrast to the inhibitory effect of ouabain due to binding to specific sodium–potassium-activated ATPase sites located on the apical surface of the choroidal epithelium (Quinton *et al.*, 1973). More recently, Lindvall-Axelsson *et al.* (1989) have shown that betamethasone also inhibits this enzyme by some 43% in a dose that inhibits CSF formation by only 31% and 24% in rabbit and rat, respectively. These authors, on the basis of the brief presentation of Fiehn and Seiler (1977), suggest that this *in vitro* reduction of ATPase may also be mediated by interference with membrane phospholipids and cholesterol (Lindvall-Axelsson *et al.*, 1989).

These experiments again support the important role of the sodium pump and Na^+/K^+-activated ATPase in the apical choroidal membrane in the elaboration of CSF. The notion of Reulen and Kreysch (1973) that the lowering of intracranial pressure in cerebral oedema is due not only to the effect of steroids on the permeability of the cerebral capillary but also to the movement of oedema fluid into the ventricles after the inhibition of CSF formation is supported by the observations presented here.

CHOROID PLEXUS BENZODIAZEPINE RECEPTORS AND SECRETION

During the past decade two types of benzodiazepine receptors have been described within the central nervous system. Of most importance, from a clinical standpoint, are the central binding receptors through which pharmacological agents (e.g. diazepam) produce their effect on nervous system function. Also of some interest was the recognition of a second type of benzodiazepine receptor (peripheral binding receptor) in heart, kidney and areas of the nervous system that are generally but not exclusively outside the blood–brain barrier (e.g. choroidal and ventricular ependyma: Richards and Mohler, 1984; Yamamura *et al.*, 1985). The high density of these peripheral benzodiazepine receptors in choroid plexus, as demonstrated by the uptake of a radiolabelled analogue of diazepam (4'-chloro analogue: Ro5-4864), suggested to us a possible role of these receptors in controlling the rate of CSF formation (Williams *et al.*, 1990). The kinetics of Ro5-4864 binding was carried out on pooled rabbit choroid plexus and brain tissue using [^{3}H]-Ro5-4864 in concentrations varying between 0.25 nM and 16 nM. The effect of this ligand on CSF formation was accomplished in the rabbit, using a ventriculocisternal perfusion system, as previously described in the dexamethasone study. The Ro5-4864 ligand was added to the ventricular perfusate in a range of concentrations between 10^{-8} M and 10^{-4} M. In addition, the effect of this benzodiazepine ligand on plexus ATPase and adenosine 3',5'-cyclic monophosphate (cAMP) was carried out, using a colorimetric and double-antibody radioimmunoassay, respectively.

The results of the kinetic experiments are shown in Figure 6.3.

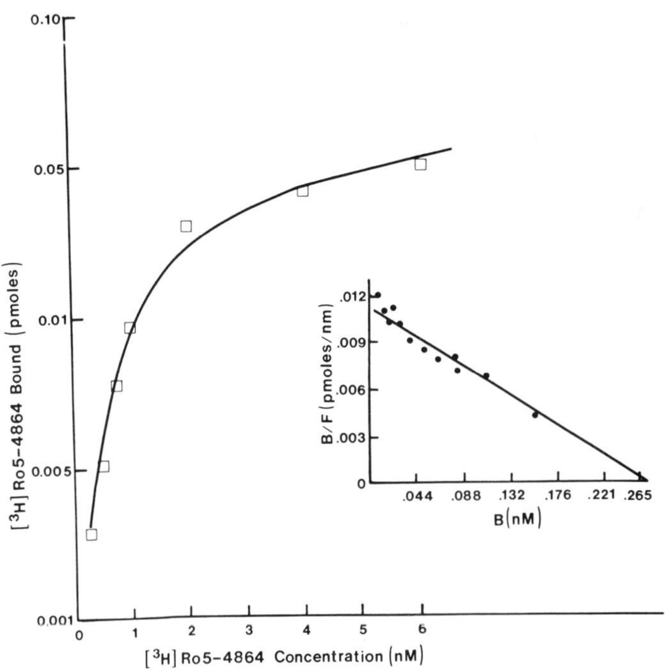

Figure 6.3 Binding of Ro5-4864 to plexus membrane. The inset represents a Scatchard plot of bound (B) against the ratio bound/free ligand (B/F). These values, corrected for sample volume and protein, were used to calculate ligand affinity (K_d) and receptor density (B_{max}), as presented in Table 6.3. From Williams *et al.* (1990)

The linearity of the Scatchard plot in this figure (inset) is compatible with a single class of choroid plexus binding sites. The total receptor density (B_{max}) can be measured from the x-axis intercept, while the slope represents $-1/K_d$. The values in brain and plexus for B_{max} and K_d are shown in Table 6.3.

Table 6.3 Peripheral benzodiazepine (Ro5-4864) receptors in brain and choroid plexus[a,d]

Tissue	K_d (nM)[b]	B_{max} (pmol/mg)[c]
Cerebral cortex	15.1	0.5
Choroid plexus	16.1	2.3

[a] Corrected for volume and protein.
[b] K_d represents affinity of the ligand.
[c] B_{max} represents total receptor density.
[d] This table from Table 1 in Weiss and Nulsen (1970).

It can be seen that the affinity of this benzodiazepine ligand for plexus tissue is relatively high and that the density of these peripheral binding sites (B_{max}) is 4.6 times greater than that found in cerebral cortex. When Ro5-4864 is added to the fluid perfusing the ventricular system, there is a maximum inhibition of CSF production by 48.5% at a concentration of 10^{-4} M, although some effect was still observed at 10^{-8} M (Table 6.4).

Table 6.4 Inhibition of CSF formation by Ro5-4864[a]

% Inhibition	10^{-4} M	10^{-5} M	10^{-6} M	10^{-7} M	10^{-8} M
No. of animals	9	4	4	4	4
Mean ± SEM	48.53	37.80	26.65	24.88	10.73
	2.13	7.02	3.71	6.16	3.89

[a] Data from Williams *et al.* (1990).

The incubation of choroid plexus tissue with 10^{-4} M Ro5-4864 did not appear to affect the level of either ATPase or cAMP as compared with control values of 3.79 ± 0.8 μmol per 10 min per mg protein and 68.0 ± 8.0 μmol per 10 min per mg protein, respectively.

The results of these experiments support the view that the high density of peripheral-type benzodiazepine receptors in choroid plexus modulate the rate of CSF production within the ventricular system. The mechanism involved in this inhibition is still undefined but does not appear to be associated with the ATPase-controlled sodium pumps on the apical surface of the choroidal cell or the cAMP second messenger system (Williams *et al.*, 1990).

SUMMARY

The three pharmacological agents presented here demonstrate that total inhibition of CSF formation is not possible with only one agent. In the case of cardiac glycosides, total Na^+/K^+-activated ATPase inhibition resulted in a less than total inhibition of fluid formation. The non-specific effect of steroids on choroidal membrane ATPase also led to a sub-total decrease in CSF production. This indicates that there is more than one mechanism involved in intraventricular fluid formation be it a choroidal or non-choroidal source. This view is also supported by the more recent demonstration of specific peripheral-type benzodiazepine receptors in choroid plexus tissue and the significant but incomplete modulation of the secretory process with a benzodiazepine analogue (Williams *et al.*, 1990). It can be predicted from these studies that multiple-drug therapy will be required to totally control CSF formation.

REFERENCES

Amano, Y. (1969). The cerebrospinal fluid production rate in the experimentally induced edematous brain and influence of dexamethasone upon it. *Nagoya J. Med. Sci.*, **31**, 427–441

Cham, B. E., Sadowski, B., O'Hagan, J. M., *et al.* (1980). High performance liquid chromatography assay of dexamethasone in plasma and tissue. *Therapeutic Drug Monitoring*, **2**, 373–377

Davson, H. and Pollay, M. (1963). Influence of various drugs on the transport of ^{131}I and PAH across the cerebrospinal fluid–blood barrier. *J. Physiol. (London)*, **167**, 239–246

Davson, H. and Segal, M. B. (1970). The effects of some inhibitors on the turnover of [22]Na in the cerebrospinal fluid and brain. *J. Physiol. (London)*, **209**, 131–153

Fiehn, W. and Seiler, D. (1977). (Na^+,K^+)-ATPase activity and sterol composition of membranes. *Naunyn-Schmiedeberg's Arch. Pharmakol.*, **297**, S21

Garcia-Bengochea, F. (1965). Cortisone and the cerebral spinal fluid of a non-castrated cat. *Am. Surg.*, **31**, 123–125

Lanzetta, P. A., Alvarez, L. J., Reinach, P. S., *et al.* (1979). An improved assay for nanomole amounts of inorganic phosphate. *Anal. Biochem.*, **100**, 95–97

Lindvall-Axelsson, M., Hedner, P. and Owman, C. (1989). Corticosteroid action on choroid plexus: reduction in Na^+-K^+-ATPase activity, choline transport capacity, and rate of CSF formation. *Exp. Brain Res.*, **77**, 605–610

Long, D. M., Hartmann, J. F. and French, L. A. (1966). The response of experimental cerebral edema to glucosteroid administration. *J. Neurosurg.*, **24**, 843–854

Martins, A. N., Newby, N. and Doyle, T. F. (1977). Sources of error in measuring cerebrospinal fluid formation by ventriculocisternal perfusion. *J. Neurol. Neurosurg. Psychiatr.*, **40**, 645–650

Martins, A. N., Ramirez, A., Solomon, L. S., *et al.* (1977). The effect of dexamethasone on the rate of formation of cerebrospinal fluid in the monkey. *J. Neurosurg.*, **41**, 550–554

Mayman, C. I. (1972). Inhibitory effect of dexamethasone on sodium-potassium activated adenosine triphosphatase of choroid plexus in cat and rabbit [abstract]. *Fed. Proc.*, **31**, 591

Mayman, C. I. and Tijerina, M. L. (1978). Inhibitory effect of steroids in Na^+-K^+ ATPase of human choroid plexus [abstract]. *Neurology*, **31**, 368

Pappenheimer, J. R., Heisey, S. R. and Jordan, E. F. (1961). Active transport of diodrast and phenosulfonthalein from cerebrospinal fluid to blood. *Am. J. Physiol.*, **200**, 1–10

Pollay, M. and Curl, F. (1967). Secretion of cerebrospinal fluid by the ventricular ependyma of the rabbit. *Am. J. Physiol.*, **213**, 1031–1038

Pollay, M., Hisey, B., Reynolds, E., Tomkins, P., Stevens, A. and Smith, R. (1985). Choroid plexus Na^+/K^+-activated adenosine triphosphatase and cerebrospinal fluid formation. *Neurosurgery*, **17**(5), 768–772

Quinton, P. M., Wright, E. M. and Tormey, J. McD. (1973). Localization of sodium pumps in the choroid plexus epithelium. *J. Cell Biol.*, **58**, 724–730

Reulen, H. J. and Kreysch, H. G. (1973). Measurement of brain tissue pressure in cold induced cerebral edema. *Acta Neurochir.*, **29**, 29–40

Richards, J. G. and Mohler, H. (1984). Benzodiazepine receptors. *Neuropharmacology*, **23**, 233–242

Sato, O. (1967). The effect of dexamethasone on cerebrospinal fluid production rate in the dog. *Brain Nerve (Tokyo)*, **19**, 485–492

Sato, O., Hara, M., Asai, T., *et al.* (1973). The effect of dexamethasone phosphate on the production rate of cerebrospinal fluid in the spinal subarachnoid space of dogs. *J. Neurosurg.*, **29**, 480–484

Vela, A. R., Carey, M. E. and Thompson, B. M. (1979). Further data on the acute effects of intravenous steroids on canine CSF secretion and absorption. *J. Neurosurg.*, **50**, 477–482

Weiss, M. H. and Nulsen, F. E. (1970). The effect of glucocorticoids on CSF flow in dogs. *J. Neurosurg.*, **32**, 452–458

Williams, G., Pollay, M., Seale, T., Hisey, B. and Roberts, A. (1990). Benzodiazepine receptors and cerebrospinal fluid formation. *J. Neurosurg.*, **72**, 759–762

Yamamura, H. I., Gehlert, D. R., Gee, K. W., *et al.* (1985). Specific high affinity [^{3}H]Ro5-4864 benzodiazepine binding sites in the brain and periphery. *Prog. Clin. Biol. Res.*, **192**, 187–196

7

Structural, Ultrastructural and Functional Correlations among Local Capillary Systems within the Brain

Joseph Fenstermacher, Hiroyuki Nakata, Atsushi Tajima, Mao-Hsiung Yen, Virgil Acuff and Kurt Gruber

INTRODUCTION

The cerebrovascular system supports the activities of brain cells by delivering essential substrates such as oxygen and glucose to brain tissue and removing the by-products of cerebral metabolism such as carbon dioxide and hydrogen ions. Cerebral blood vessels also carry blood-borne hormones and homoeostatic information (e.g. plasma osmolality) to the brain and distribute neurally synthesized humoral agents from brain to systemic sites.

Within the central nervous system there are broad differences in local functions such as neuronal activity, cellular organization, afferent and efferent connections, neural transmitters, endocrine activity and glucose utilization. In view of these variations and the published interarea variations in local cerebral blood flow, large differences in capillary bed structures and function most probably exist. The purpose of this study was the quantification of the volume of circulating labelled red cells, albumin and blood in parenchymal microvessels, local blood flow, mean capillary diameter, capillary luminal volume and surface area, and frequency of endothelial cell vesicles, junctions, fenestrae and mitochondria. These measurements were made in up to eleven brain areas, which included grey matter structures, small hypothalamic nuclei, white matter and several circumventricular organs. The results were compared among brain areas, and correlation between measured variables were examined.

METHODS

Radiotracer Studies

Male Sprague–Dawley rats weighing 280–350 g were lightly anaesthetized with a mixture of 1.0–1.5% halothane, 30% oxygen and 70% nitrous oxide. Polyethylene catheters were inserted into one femoral vein and both femoral arteries. The wounds were infiltrated with a local anaesthetic. The rat was allowed to recover from surgery and anaesthesia for at least 2 h. During the recovery and experimental periods, arterial blood pressure, blood gases and pH, haematocrit and rectal temperature were measured.

The distribution of various radioactive tracers was determined in a separate set of animals, using previously published techniques. As each method required measurement of radioactivity in arterial blood during the experiment, blood samples were drawn from a femoral artery at prescribed times throughout each study and their levels of radioactivity were assessed by scintillation or gamma counting.

At the conclusion of all experiments, the brain was removed, frozen and cut into a series of 20-μm-thick coronal sections in a cryostat at $-17\,°C$. Adjacent sections were collected and stained with cresyl violet for histological identification of individual brain structures. The sections for autoradiography were placed in cassettes with film for 5–21 days, the length of time depending on the tracer used. Tissue radioactivity was assessed by quantitative autoradiography. The following measurements were made within a number of brain structures: (1) the rate of local cerebral blood flow (LCBF), using [^{14}C]-iodoantipyrine (IAP; Sakurada *et al.*, 1978); (2) the blood-to-brain transfer constant (K_1) of alpha-aminoisobutyric acid (AIB; Blasberg *et al.*, 1983); (3) the distribution volume of ^{51}Cr-labelled red blood cells in parenchymal microvessels (V_r; Gross *et al.*, 1987); and (4) the distribution volume of [^{125}I]-albumin in parenchymal microvessels (V_p; Gross *et al.*, 1987). These data were combined to yield various physiological functions: (1) the distribution volume of radiolabelled blood in parenchymal microvessels ($V_b = V_r + V_p$); (2) the haematocrit in the parenchymal micovessels (Hct $= V_r/V_b$); (3) the rate of local plasma flow (LCPF = LCBF × arterial haematocrit); and (4) the permeability–surface area product for AIB (PS $= -$LCPF ln $1 - K_1$/LCPF).

Morphometric Studies

Male Sprague–Dawley rats weighing 250–300 g were anaesthetized with sodium pentobarbitol (40 mg/kg, i.p.). The upper body was initially perfused with heparinized buffered saline through a cannula in the left ventricle. About 30 s later a second perfusion through the same cannula was made with a solution of 2% paraformaldehyde and 2% glutaraldehyde in 0.1 M phosphate buffer at pH 7.4 and 37 °C. The brain was removed and stored for at least 24 h in cold 2% glutaraldehyde in 0.1 M buffer.

A set of 200-μm-thick coronal sections were cut the following day with a vibratome. Samples of specific brain areas were cut from the vibratomed sections, placed in a 0.1 M phosphate buffer and postfixed in buffered 1% osmium tetroxide. These samples were then dehydrated through a graded series of alcohols and propylene oxide and embedded in epoxy resin ('plastic'). From each tissue specimen, 2-μm-thick sections were cut with glass knives and stained with 1% toluidine blue.

For morphometric analysis at the light microscopic level, photographs of the sections were taken with a Nikon Optiphot and enlarged to a final magnification of ×400 or ×800. An area for analysis was outlined on the micrographs, and all cross-sectioned microvessels within that area with short-axis diameters less than 7.5 μm were counted as 'capillaries'. A Micro-Plan II (DonSanto Corp., Natick, Mass., USA) was used to measure the total tissue area, the number and area of capillary profiles within the measured area, and capillary diameter. From these measurements, the capillary density (N_a), the volume fraction (V_v), and the surface area of capillaries (S_v) and the mean capillary diameter (D) were calculated (Lin *et al.*, in press).

Morphometric analysis at the ultrastructural level was performed on the same fixed-tissue specimens as were used for light microscopy. Ultrathin silver sections (about 75 nm thick) were cut with a diamond knife and stained with uranyl acetate and lead citrate. Round or nearly round capillary profiles were located with a JEOL 100 CX transmission electron microscope, and micrographs of endothelial cells were obtained from these capillaries. These micrographs were subsequently enlarged to a final magnification of ×25 000. The numbers of cell junctions, mitochondria, fenestrae and vesicular profiles per capillary profile were quantificated.

Brain Structures

For the quantitative autoradiographic studies, the following brain structures were analysed: inferior colliculus (IC), sensorimotor cortex (SMC), ventromedial nucleus of the hypothalamus (VMH), supraoptic nucleus of the hypothalamus (SON), genu of the corpus callosum (GCC), globus pallidus (GP), ventricular surface of the caudate putamen (CPV), central part of the caudate putamen (CPC), subfornical organ (SFO), median eminence (ME) and neural lobe of the pituitary (NL). Seven of these same brain structures plus the paraventricular nucleus of the hypothalamus (PVN) were analysed for the morphometric studies.

The properties of the capillaries differ among these brain areas. The subfornical organ, median eminence and pituitary neural lobe have fenestrated, highly permeable capillaries, whereas the capillaries in the other nine brain areas (IC, SMC, GCC, VMH, SON, PVN, GP, CPV and CPC) are unfenestrated and transport-restrictive—namely, they are typical blood–brain barrier (BBB) capillaries. Functionally, the SFO, ME and NL are part of the neuroendocrine system and are included in a peculiar group of brain structures called the circumventricular organs (CVO).

Data Presentation

The data are reported as means ± SE. The number of rats per group ranges from 5 to 13.

RESULTS AND DISCUSSION

The physiological condition of the rats used in this study was normal for awake, restrained animals. For example, the mean (±SE) values for the 13 rats in the ^{51}Cr-labelled red cell experiments were: blood glucose = 180 ± 12 mg/dl; mean arterial blood pressure = 127 ± 2.5 mmHg; arterial haematocrit = 0.49 ± 0.006; plasma osmolality = 291 ± 4 mOsm/kg water; arterial pH = 7.42 ± 0.006; arterial $P_{CO_2} = 43.6 \pm 0.8$ mmHg; and arterial $P_{O_2} = 92.9 \pm 3.4$ mmHg.

Distribution Volumes

The distribution of ^{51}Cr-tagged red cells (RBC) was measured at five times ranging from 15 s to 5 min after completing intravenous administration, which takes 20–30 s for the volumes injected in this study. The ^{51}Cr]-RBC distribution volumes (V_r) were constant for each brain area from 15 s to 5 min. This indicates that the number of microvessels labelled by ^{51}Cr]-RBCs did not increase over time, suggests that the same population of microvessels were perfused for this 5 min period, and argues against intermittency of cerebral capillary perfusion.

Because V_r did not increase over time, a mean value was determined from all of the data for each brain area. The mean ^{51}Cr]-RBC distribution volumes varied twentyfold among brain areas (Table 7.1); V_r was lowest for the corpus callosum, a white matter structure, and highest for the neural lobe, one of the circumventricular organs. Among the grey matter areas, V_r was smallest in the three basal ganglia areas (GP, CPV and CPC) and the ventromedial nucleus of the hypothalamus, intermediate in the sensorimotor cortex, and highest in the inferior colliculus and the supraoptic nucleus, which is also in the hypothalamus.

Table 7.1 Microvascular distribution volumes, haematocrits and rates of blood flow (LCBF) within eleven brain areas[a]

Brain areas	RBC volume (μl/g)	RISA volume (μl/g)	Blood volume (μl/g)	Haematocrit (%)	LCBF (ml g^{-1} min^{-1})
GCC	1.1±0.1 (13)	3.7±0.2 (13)	4.8±0.2	23±1	53±2 (9)[b]
GP	2.9±0.1 (13)	5.0±0.3 (13)	7.9±0.3	36±2	83±6 (6)[b]
VMH	2.9±0.1 (13)	5.0±0.3 (13)	7.9±0.4	37±2	125±9 (5)[c]
CPV	3.2±0.2 (13)	5.4±0.3 (13)	8.6±0.3	38±2	142±9 (9)[b]
CPC	3.3±0.2 (13)	5.8±0.3 (13)	9.1±0.3	36±2	122±8 (9)[b]
SMC	4.1±0.2 (13)	6.5±0.4 (13)	10.6±0.5	39±2	200±27 (9)[b]
IC	5.8±0.3 (12)	7.0±0.4 (12)	12.8±0.5	45±2	218±21 (9)[b]
SON	6.5±0.4 (13)	13.1±0.4 (11)	19.6±0.5	33±1	328±35 (5)[c]
SFO	10.2±1.7 (12)	21.5±3.1 (8)	31.7±3.6	32±5	108±14 (6)[b]
ME	18.7±1.9 (13)	23.3±1.5 (13)	41.9±2.4	45±2	577±33 (5)[c]
NL	22.6±2.9 (13)	46.0±5.0 (13)	68.6±5.0	33±3	886±11 (5)[c]

[a] Values are means ± SE, with number of animals in parentheses.
[b] Data are from Tajima *et al.* (submitted).
[c] Data are from Bryan *et al.* (1988).

Large [^{51}Cr]-RBC volumes were found in the three circumventricular organs studied (SFO, ME and NL) and agree with the known high vascularity of these structures (Gross and Weindl, 1987). Markedly smaller distribution volumes are obtained in these three brain areas when ^{55}Fe-labelled red cells are used (Lin *et al.*, 1990). Specifically, the mean [^{55}Fe]-RBC distribution volumes in the SFO, ME and NL are 3.9, 3.6 and 10.2 µl/g, respectively, and are 20–50% of the [^{51}Cr]-RBC values in Table 7.1. As discussed by Lin *et al.*, the preparation of ^{51}Cr-labelled red cells alters their membrane properties, which probably affects the distribution of those erythrocytes in at least some brain areas.

In the present study the animals were killed by decapitation, and the brains were removed from the skull prior to freezing. The observations of Weiss and Edelman (1976) and Lin *et al.* (1990) indicate that in rabbits and rats, respectively, all or virtually all parenchymal microvessels with internal diameters less than 100 µm retain red cells following decapitation and brain removal. Accordingly, the red cell distribution volumes listed in Table 7.1 are for a combined population of parenchymal microvessels, including arterioles, capillaries and venules.

The distribution volumes of [^{125}I]-serum albumin (RISA) were measured at five times after completing intravenous administration, which was accomplished by a 10 s infusion. From 15 s, the earliest time of sampling, to 5 min, the longest circulation period, the RISA distribution volumes were constant in all 11 brain areas. This observation is consistent with the constancy of V_r reported above and suggests that the microvessels were not intermittently perfused during the period of 15 s–5 min under these conditions.

Because of the constancy of V_p over time, mean values were calculated from all the data of each brain area. A twelvefold range of RISA distribution volumes were found for this set of brain areas (Table 7.1). The lowest V_p was measured in the corpus callosum. With the exception of the SON, the RISA spaces in the grey matter areas were between 5.0 and 7.0 µl/g. Among the CVOs, the distribution volumes of RISA were large—and similar—in the subfornical organ and median eminence and very large in the neural lobe.

The volume of radiolabelled blood in parenchymal microvessels, V_b, is the sum of V_r and V_p and ranges from about 5 µl/g (GCC) to nearly 70 µl/g (NL; Table 7.1). The difference in blood volumes between the two hypothalamic nuclei, the VMH and SON, was sizable. The large labelled blood volumes in the three neuroendocrine structures, the SFO, ME and NL, are consistent with the importance of the vascular system in their function. Incidentally, the supraoptic nucleus, which also has a large blood volume, may be involved in neuroendocrine regulation of body fluid volumes and osmolality, as are the SFO, ME and NL (Gross *et al.*, 1986b; Gross and Weindl, 1987).

As previously indicated, the arterial haematocrit in these animals was about 49%. The microvessel haematocrits in most of these 11 brain areas (Table 7.1) were significantly less than the arterial haematocrit. Differences in arterial and cerebral microvascular haematocrits have been repeatedly reported (e.g. Gross *et al.*, 1986a,b). The microvascular haematocrits measured with [^{55}Fe]-RBC data are fairly similar to those derived from [^{51}Cr]-RBC data for most brain areas. For the circumventricular organs, however, much smaller haematocrits are obtained with [^{55}Fe]-RBCs (13.5% for the ME, 15.3% for the SFO and 18.2% for the NL: Lin *et al.*, 1990). The low haematocrits obtained with [^{55}Fe]-RBCs probably more closely approximate the 'unlabelled red cell' or normal haematocrit in the microvessels in these three brain areas, since the ^{51}Cr procedure for labelling red cells alters membrane charges and other membrane properties as indicated above.

To have mass balance (i.e. inflow to equal outflow) in the system, the velocity of red cell flow must be greater than the velocity of plasma flow when the haematocrit within the microvessels is less than the central arterial haematocrit (the Fahreaus effect). The low microvessel haematocrits in brain areas such as the corpus callosum and supraoptic nucleus (Table 7.1) suggest that red cells pass through at least some cerebral microvascular systems approximately twice as fast as plasma constituents such as albumin.

Local Cerebral Blood Flow

Local cerebral blood flow (LCBF) is usually measured with highly permeable markers such as iodoantipyrine (IAP). Recent evidence suggests that IAP can move substantial distances within the tissue after crossing the capillary wall (Jay *et al.*, 1988) and result in poor estimates of LCBF (Tajima *et al.*, submitted). In particular, IAP appears to underestimate LCBF substantially in small brain areas with high flow rates that are adjacent to the cerebrospinal fluid (CSF) such as the supraoptic and paraventricular nuclei of the hypothalamus (Bryan *et al.*, 1988). For this reason the LCBF values determined by Bryan *et al.* with [^{14}C]-isopropyliodoamphetamine (IPIA) are 'reported' herein for the VMH, SON, ME and NL (Table 7.1). The subfornical organ is a small structure that forms part of the roof of the third ventricle. Diffusional loss of IAP to the CSF could be substantial for this structure, and LCBF in the subfornical organ may be two or more times larger than that measured with IAP and listed in Table 7.1.

Be that as it may, local cerebral blood flow extends from a low of 53 ml 100 g^{-1} min^{-1} in the corpus callosum to a high of 886 ml 100 g^{-1} min^{-1} in the neural lobe (Table 7.1). This sixteenfold range of LCBFs correlates well with the variations in local microvascular blood volumes (Table 7.1; Figure 7.1). Of the 11 brain areas studied, the V_b–LCBF relationship is clearly different for only the subfornical organ, a structure in which blood flow may be severely underestimated when measured with iodoantipyrine (see above). If the flow rate in the SFO is around 300 ml, then the correlation between V_b and LCBF is very tight and suggests that the major cause of the differences in flow rates among both circumventricular organs and non-CVO brain areas is the variation in the number of perfused microvessels.

Capillary Bed Structure

The sole criterion used to select capillary profiles was the length of the short or minor axis of the lumen. All profiles with 'short-axis diameters' equal to or less than 7.5 µm were counted as capillaries. The short axis is easy to determine from round or oval profiles. However, many microvessel profiles have complex shapes and estimating the site and length of the short axis of such profiles is rather trickier. The procedure for doing this is given by Lin *et al*. (1990).

The mean diameters of the capillaries ranged from 4.0 µm in the supraoptic nucleus to 5.3 µm in the corpus callosum (Table 7.2). The mean diameter was significantly ($p<0.05$) larger in the GCC, VMH, SMC and IC than in the SON, PVN, SFO and NL. All other things being equal, the resistance to fluid flow through the capillaries in the latter four brain areas would be expected to be relatively high, owing to the small bore of those microvessels. The similarity in mean diameters between the SON and PVN, two hypothalamic structures thought to be involved in osmo-reception and vasopressin synthesis, and the SFO and NL, two neuroen-docrine areas, suggests that small capillary diameter within the brain may be related to local involvement in the regulation of fluid and electrolyte balance in the body.

Within the brain, capillaries are curved, tortuous structures (Yoshida and Ikuta, 1984), and the same capillary may appear two or more times in the area of measurement on the histological section. The frequency of capillary profiles (N_a) is, therefore, equal to or greater than the number of individual capillaries sectioned (N_c), and probably does not equal capillary density in any of the brain areas studied. Capillary profile frequency was lowest in the corpus callosum and ventromedial nucleus, where mean capillary diameter is large, and highest in the paraventricular nucleus and neural lobe, where mean capillary diameter is small (Table 7.2). At the moment no functional reason for this 'reciprocal' relationship is clear.

The volume fraction (V_v) approximates the volume of blood that could be contained in capillaries and ranges nearly eightfold among these brain

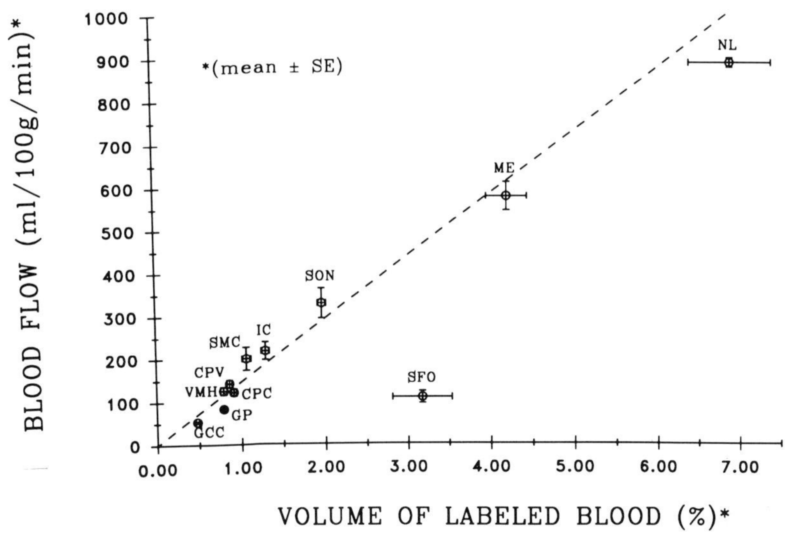

Figure 7.1 The relationship between local cerebral blood flow and volume of labelled blood for eleven brain areas. All values are taken from Table 7.1. The abbreviations are listed under Methods. The dashed line was drawn by eye, starting at the origin

areas (Table 7.2). The volume of radiolabelled blood in parenchymal microvessels (Table 7.1) is approximately equal to V_v in the GCC and VMH, less than V_v in the SMC, IC and SON, and greater than V_v in the SFO and NL. Clearly, labelled red cells and serum albumin are retained in microvessels larger than capillaries in the SFO and NL. The cause of the low volumes of labelled blood in the SMC, IC and SON may be perfusion and labelling of only a fraction of the capillaries in these structures during the experimental period.

Capillary surface area varied more than ninefold among those eight

Table 7.2 Average morphometric parameters of local cerebral capillary systems in eight brain areas of Sprague–Dawley rats[a]

Area	N	$D(\mu m)$	$N_a(No./mm^2)$	$V_v(\%)$	$S_v(mm^2/mm^3)$
GCC	5	5.3 ± 0.05	150 ± 3.6	0.53 ± 0.02	4.14 ± 0.16
VMH	5	5.0 ± 0.07	231 ± 7	0.81 ± 0.02	6.56 ± 0.17
SMC	5	5.1 ± 0.02	327 ± 5	1.24 ± 0.05	9.74 ± 0.45
IC	5	4.8 ± 0.09	500 ± 17	1.76 ± 0.03	14.8 ± 0.52
SON	5	4.0 ± 0.11	983 ± 52	3.04 ± 0.06	30.0 ± 0.59
PVN	6	4.1 ± 0.05	1269 ± 29	3.99 ± 0.12	38.9 ± 1.04
SFO	5	4.1 ± 0.06	578 ± 25	1.66 ± 0.12	16.2 ± 0.93
NL	8	4.1 ± 0.9	1072 ± 50	3.53 ± 0.19	34.6 ± 1.49

[a] Values are means $\pm$ SE. Most of these data have been previously published (Gross *et al.*, 1986a). The number of animals has been increased in some cases. All values have been corrected for a previous analytical error.

brain areas (Table 7.2) and was very large in the supraoptic nucleus, paraventricular nucleus and neural lobe. Capillary surface area would seem to be a function of local need for blood–brain exchange of materials such as glucose and amino acid and for exposure of endothelial cell receptors to blood-borne ligands such as peptides. This broad variation in brain capillary surface area must, therefore, be related to and be regulated by local activities. Rapid but modest regulation of capillary surface area could be obtained by varying the number of perfused microvessels, whereas long-term regulation might involve remodelling of capillary beds—namely, microvascular angiogenesis.

Capillary Endothelial Cell Ultrastructure

The transfer of solutes from blood to tissue is known to be much higher in circumventricular organs than in other brain areas. For example, the PS products of alpha-aminoisobutyric acid are 200–300 times higher in the subfornical organ and neural lobes than in the GCC, SMC and SON (Gross *et al.*, 1986a,b). Morphological features that may be correlated and involved with these marked differences in cerebral capillary permeability include endothelial cell vesicles and fenestrae.

It is commonly stated that the BBB is structurally characterized by a 'paucity of vesicles' and that the frequency of endothelial cell vesicles is very great in brain areas with highly permeable capillaries and in conditions where the BBB is open or leaky. These observations imply that vesicles are the facilitators of transendothelial cell movement in permeable cerebral capillaries. Ultrastructural morphometry indicated a threefold variation in vesicular frequency within endothelial cells of tight (BBB) capillaries and fairly similar frequency between the capillaries of the supraoptic nucleus and neural lobe, two brain areas with markedly different capillary permeabilities (Figure 7.2A). These data show that drawing conclusions on capillary permeability and mechanism of transcapillary passage based on vesicular frequency can be very misleading and frankly erroneous.

Lipid-insoluble materials are thought to cross most capillary walls by way of the intercellular clefts or junctions. Differences in capillary permeability might, therefore, correlate well with junctional frequency. Cerebral capillary data do not support this, since junctional frequency in the highly permeable capillaries of the NL and SFO is fairly similar to that in the slightly permeable capillaries of the GCC, ICC, SMC, VMH and SON (Figure 7.2B). Recently Crone (1986) and Frokjaer-Jensen (1990) have postulated that capillary permeability is a function of the complexity of the junctions and the percentage of the junctions available for polar substance movement. For instance, the differences in permeability among

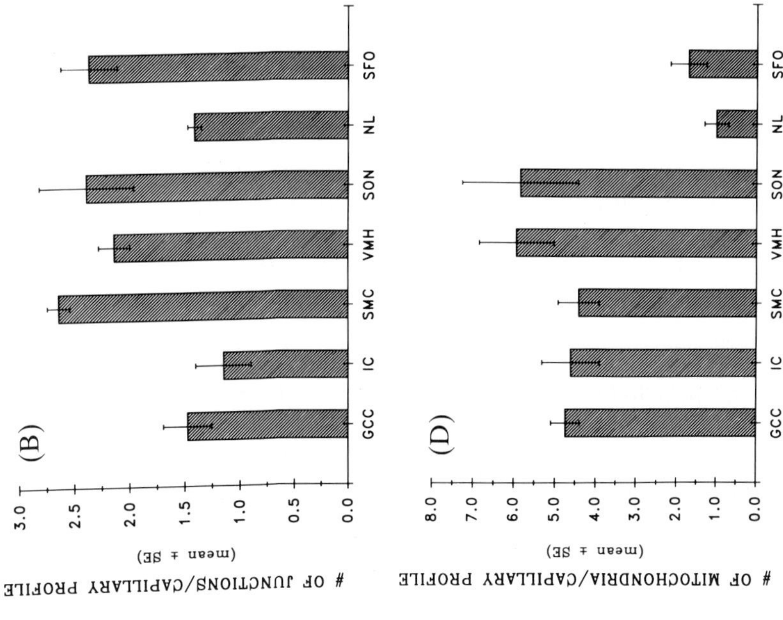

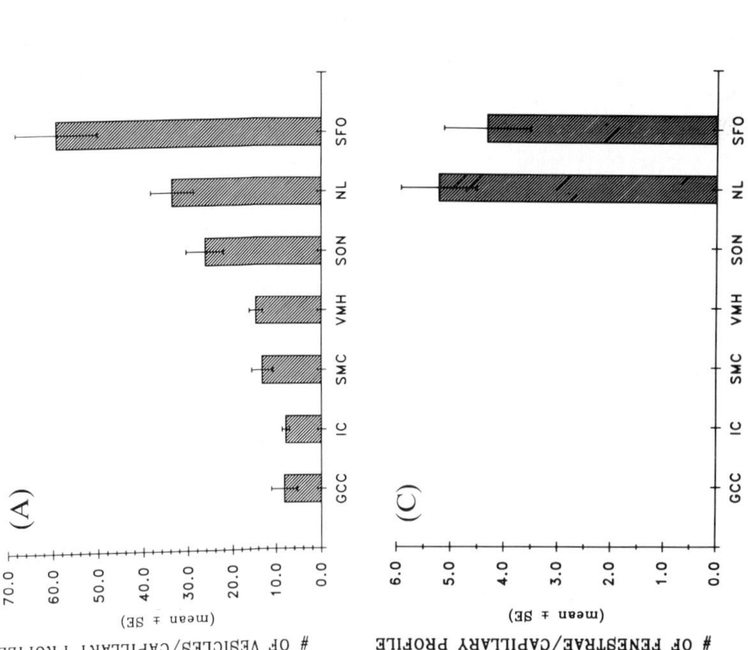

Figure 7.2 The frequency (No./profile) of various ultrastructural features of capillary endothelial cells from seven areas of the rat brain. (A) Frequency of vesicular profiles per capillary cross-section. (B) Frequency of intercellular junctions or clefts per capillary profile. (C) Frequency of fenestrae per capillary cross-section. (D) Frequency of mitochondrial profiles per capillary profile. The abbreviations are listed under Methods. Some of these data were previously published (Gross *et al.*, 1986a,b, 1987)

mesenteric, skeletal muscle and brain capillaries can be explained if 85%, 12% and 0.1% of the intercellular cleft, respectively, is open for transfer. Testing of the suggestions of Crone and Frokjaer-Jensen cannot be made with our ultrastructural data.

The endothelial cells in brain areas with tight capillary walls are unfenestrated, whereas those in the highly permeable capillaries of the neural lobe and subfornical organ have 4–5 fenestrae per capillary cross-section (Figure 7.2C). The correlation between capillary permeability and frequency of endothelial cell fenestrations is, therefore, quite good.

Oldendorf *et al.* (1977) observed that brain capillaries contain a relatively large number of mitochondria and suggested that some of this potential work capacity could be employed in ionic and molecular transfer across the BBB. Our ultrastructural analysis indicates that the frequency of mitochondrial profiles is very similar in endothelial cells from five brain areas with tight capillary walls (Figure 7.2D) and agrees with the data of Oldendorf *et al.* In the neural lobe and subfornical organ, the frequency of mitochondrial profiles is much lower than in other brain areas. This indicates that the activities of the endothelial cells require a greater energy supply or work capability in brain areas such as the hypothalamus (VMH and SON) than do those in the neural lobe and subfornical organ. Since there is probably little carrier-mediated transfer of solutes such as D-glucose and L-phenylalanine and pumping of ions such as potassium across capillary walls in the neural lobe and subfornical organ, the differences in mitochondrial frequency and apparent work capacity between CVO and non-CVO cerebral capillaries may, indeed, be related to the differences in these transport functions, as postulated by Oldendorf *et al.*

ACKNOWLEDGEMENTS

The authors thank Grace Richardson, Nadine Sposito, Susan Pettersen and Lisa Rybacki for their technical assistance, and Patricia Milligan for secretarial service. This work was supported by Grant NS 21157 from the National Institutes of Health.

REFERENCES

Blasberg, R. G., Fenstermacher, J. D. and Patlak, C. S. (1983). Transport of alpha-aminoisobutyric acid across brain capillary and cellular membranes. *J. Cereb. Blood Flow Metab.*, **3**, 8–32

Bryan, R. M., Myers, C. L. and Page, R. B. (1988). Regional neurohypophyseal and hypothalamic blood flow in rats during hypercapnia. *Am. J. Physiol.*, **255**, R295–R302

Crone, C. (1986). The blood–brain barrier: a modified tight epithelium. In Suckling, A. J., Rumsby, M. G. and Bradbury, M. W. B. (Eds), *The Blood–Brain Barrier in Health and Disease*. Ellis Horwood, Chichester, pp. 17–40

Frokjaer-Jensen, J. (1990). Anatomical correlates of capillary permeability. In Molinatti, G. M., Bar, R. S., Belfiore, F. and Porta, M. (Eds), *Endothelial Cell Function in Diabetic Microangiopathy: Problems in Methodology and Clinical Aspects*. Karger, Basel, Frontiers in Diabetes, Vol. 9, pp. 28–42

Gross, P. M., Blasberg, R. G., Fenstermacher, J. D. and Patlak, C. S. (1987). The microcirculation of rat circumventricular organs and pituitary gland. *Brain Res. Bull.*, **18**, 73–85

Gross, P. M., Sposito, N. M., Pettersen, S. E. and Fenstermacher, J. D. (1986a). Differences in function and structure of the capillary endothelium in gray matter, white matter and a circumventricular organ of rat brain. *Blood Vessels*, **23**, 261–270

Gross, P. M., Sposito, N. M., Pettersen, S. E. and Fenstermacher, J. D. (1986b). Differences in function and structure of the capillary endothelium in the supraoptic nucleus and pituitary neural lobe of rats. *Neuroendocrinology*, **44**, 401–407

Gross, P. M. and Weindl, A. (1987). Peering through the windows of the brain. *J. Cereb. Blood Flow Metab.*, **7**, 663–672

Jay, T. M., Lucignani, G., Crane, A. M., Jehle, J. and Sokoloff, L. (1988). Measurement of local cerebral blood flow with [^{14}C]iodoantipyrine in the mouse. *J. Cereb. Blood Flow Metab.*, **8**, 121–129

Lin, S.-Z., Sposito, N., Pettersen, S., Rybacki, L., McKenna, E., Pettigrew, K. and Fenstermacher, J. (1990). Cerebral capillary bed structure of normotensive and chronically hypertensive rats. *Microvasc. Res.* (in press)

Oldendorf, W. H., Cornford, M. E. and Braun, W. J. (1977). The large apparent work capability of the blood–brain barrier: a study of the mitochondrial content of capillary endothelial cells in brain and other tissues of the rat. *Ann. Neurol.*, **1**, 409–417

Sakurada, D., Kennedy, C., Jehle, J., Brown, J. D., Carbin, G. L. and Sokoloff, L. (1978). Measurement of local cerebral blood flow with iodo[^{14}C]antipyrine. *Am. J. Physiol.*, **234**, H59–H66

Weiss, H. R. and Edelman, N. H. (1976). Effect of hypoxia on small vessel blood content of rabbit brain. *Microvasc. Res.*, **12**, 305–315

Yoshida, Y. and Ikuta, F. (1984). Three-dimensional architecture of cerebral microvessels with a scanning electron microscope: a cerebrovascular casting method for fetal and adult rats. *J. Cereb. Blood Flow Metab.*, **4**, 290–296

8
Effects of Phenylephrine and Dopamine on Locomotor Activity and Permeability of the Blood–Brain Barrier of Mice Exposed to Lead from Birth

Floyd R. Domer

INTRODUCTION

As has been pointed out by Yule (1978), Shaffer (1978) and Millichap (1982), there is no unanimity in the classification and characterization of the minimal brain dysfunction (MBD) syndrome versus the minimal brain damage syndrome for children. The differences are probably a reflection of the fact that there are a number of potential contributing factors to the syndrome that is designated MBD.

Components that are common to most considerations of MBD are:

Hyperactivity
Short attention span
Impulsiveness
Learning disability
Age: 6–12 years
Boys more than girls

The children are most commonly classed as having MBD when they reach school age, because this is often the first time that they are placed in a highly structured environment for extended periods of time. In such a situation the fact that these children move around a great deal, are impulsive and have a short attention span makes them disruptive in the classroom and contributes to their difficulties in learning. This combination of disruptiveness and difficulty in learning brings them to the attention of their teachers in a very negative context. It is estimated that 4–10% of children under 12 years of age exhibit this syndrome (Millichap, 1982; Kalverboer, 1978). When evaluated in their teens, these children have been found to be fidgety rather than hyperactive, to be less aggressive but felt to be antisocial, and to be sad as though they were aware of low expectations for the future (Shaffer, 1978).

There have been a number of possible causes suggested for the development of the MBD syndrome. Those that seem to have the greatest support are:

Perinatal hypoxia
Meningitis
Head injury
Malnutrition
Lead

Shaffer (1978) reviewed many of these factors and concluded that a low birth weight, and therefore a more immature infant, seemed to be the most promising explanation for most cases of MBD. This might be exaggerated by malnutrition. Both meningitis and head injury may cause changes in the function of the blood–brain barrier (BBB) to alter the environment surrounding the neurons in the central nervous system (Davson *et al.*, 1987; Suckling *et al.*, 1986). The fact that the concentration of lead in the blood is elevated in children who are hyperactive has also been recognized as a possible contributor to the MBD syndrome.

Acute exposure to lead in sufficient concentrations causes encephalopathy in children. This was reviewed by Blackman (1937). The toxicity of lead had been known since ancient times. The encephalopathy was characterized by cerebral oedema, especially in the cerebral cortex and cerebellum, associated with increased intracranial pressure (Smith *et al.*, 1960). Capillaries were dilated and had swollen endothelial nuclei, while there was a proliferation of astrocytes (Pentschew, 1965). Neurons were swollen, as were glia, and the myelin was split, with a breakdown of the lamellae (Raimondo *et al.*, 1968). Subsequently there were reports using animals exposed to lead prior to and after birth that have provided information about acute and delayed effects (Pentschew and Garro, 1966; Thomas *et al.*, 1973; Press, 1977; Toews *et al.*, 1978; Johnson, 1980). Most of these studies utilized exposure of the animals to lead either *in utero* or during the neonatal period. The capillaries of the young are especially sensitive to damage caused by lead, but subsequently seem to recover (Toews *et al.*, 1978). Their number and permeability also vary (Johnson, 1980; Mitchell *et al.*, 1980). This unusual sensitivity of the young is also found in humans when the absorption and distribution of lead is considered. As reviewed by Lin-Fu (1973), children absorb much more lead from the intestinal tract and have much less of that lead sequestered in bone than do adults. This results in their having a proportionately greater amount of the lead to which they are exposed available for distribution to the soft tissues of the body, including the brain.

There have been numerous investigations of the effects of exposure to lead upon the permeability of the BBB (Lorenzo and Gewirtz, 1977; Domer and Llera, 1978, 1979; Miller *et al.*, 1979; Domer and Wolf, 1980; Kolber *et al.*, 1980; Lefauconnier *et al.*, 1980; Hertz *et al.*, 1981; Michaelson and Bradbury, 1982; Zenick *et al.*, 1982; Turnbull and Brodeur, 1984; Sandstrom *et al.*, 1985; Bradbury *et al.*, 1987). The results have been quite variable and range from a decrease in permeability (Lorenzo and Gewirtz, 1977; Domer and Llera, 1978; Kolber *et al.*, 1980; Michaelson and Bradbury, 1982; Zenick *et al.*, 1982), to no change (Hertz *et al.*, 1981) to an increase in permeability (Domer and Wolf, 1979, 1980; Miller *et al.*, 1979; Mitchell *et al.*, 1980; Sandstrom *et al.*, 1985). This variability likely reflects a number of factors that are of importance in the movement of substances across the BBB, e.g. duration of exposure, size and charge of the molecule being studied, or other drugs being employed in the study.

Children who have been diagnosed as having MBD are often treated with drugs. The compounds that are most effective in these children are:

Dextroamphetamine (Dexedrine)
Methylphenidate (Ritalin)
Magnesium pemoline (Cylert)
Caffeine
Imipramine (Tofranil)

These have been reviewed by Millichap (1982) and Pyck and Baines (1978). Each of these drugs has an interaction, either direct or indirect, with the sympathetic nervous system. The indirect aspect of their actions may well involve actions at alpha or dopaminergic receptors.

In an attempt to clarify further some of these factors, mice were exposed to lead from birth, and the effect on weight, locomotor activity and permeability of the BBB was determined when exposed to phenylephrine or dopamine.

MATERIALS AND METHODS

The method of exposure to lead was essentially that of Silbergeld and Goldberg (1973). Pregnant CD-1 mice were received from the Charles River Laboratories on the fifteenth day of gestation and were housed separately in 18×20 cm cages. They were allowed free access to food and water. Within 12 h of delivery the drinking water was changed to either a 0.5% solution of sodium acetate (control) or lead acetate (experimental). At 21 days of age the pups were weaned and the exposure to sodium or lead in the drinking water was continued. The mice were weighed at the time of weaning and weekly thereafter.

Locomotor activity was assessed at the time of weaning and weekly for the two succeeding weeks. The mice were placed individually in a box (2×2 ft) which had the bottom marked into squares (2×2 in). The number of squares crossed during a 10 min period of observation was recorded. On the day following the third evaluation of locomotor activity the drugs to be used to test the permeability of the BBB were injected intraperitoneally 15 min prior to the mice being placed in the box for evaluation of locomotor activity. This interval permitted adequate absorption of the drug. The value obtained in this observation period was compared with that obtained on the preceding day.

The function of the BBB was assessed by the intravenous injection of radioactive technetium ($[^{99m}Tc]-O_4$) 50 μCi 5 min after the administration of the drugs. One hour after the injection of the radioactivity, the mouse was anaesthetized with ether, a blood sample (0.5 ml) was obtained by cardiac puncture and saline was then infused via the left cardiac ventricle following section of the venae cavae to remove blood from the cerebral vasculature. The brain was dissected out and weighed. The weighed samples of blood and brain were counted with a Packard Auto Gamma 5650 scintillation counter. The counts were corrected for background and the ratio between counts per minute per g brain and counts per minute per ml blood was calculated.

Groups of ten mice were used for each dosage employed. The results were compared with those obtained from the control (saline) animals, using a t test (Mather, 1947).

The doses of the drugs used in this study were: saline, phenylephrine hydrochloride (0.125 or 0.25 mg/kg) or dopamine hydrochloride (2 or 4 mg/kg). The drugs were administered intraperitoneally.

RESULTS

As illustrated in Figure 8.1, the mice that had been exposed to lead from birth were significantly smaller than the mice that had been exposed to sodium from birth. The same relationships were found when the locomotor activity was determined—i.e. the animals exposed to lead were less active (Figure 8.2).

When phenylephrine was administered to the mice, it caused an increase in locomotor activity that was not significantly greater than that of the control group given saline (Figure 8.2). However, at both doses of phenylephrine the group of animals that had been exposed to lead had significantly less locomotor activity than did the animals that had been exposed to sodium.

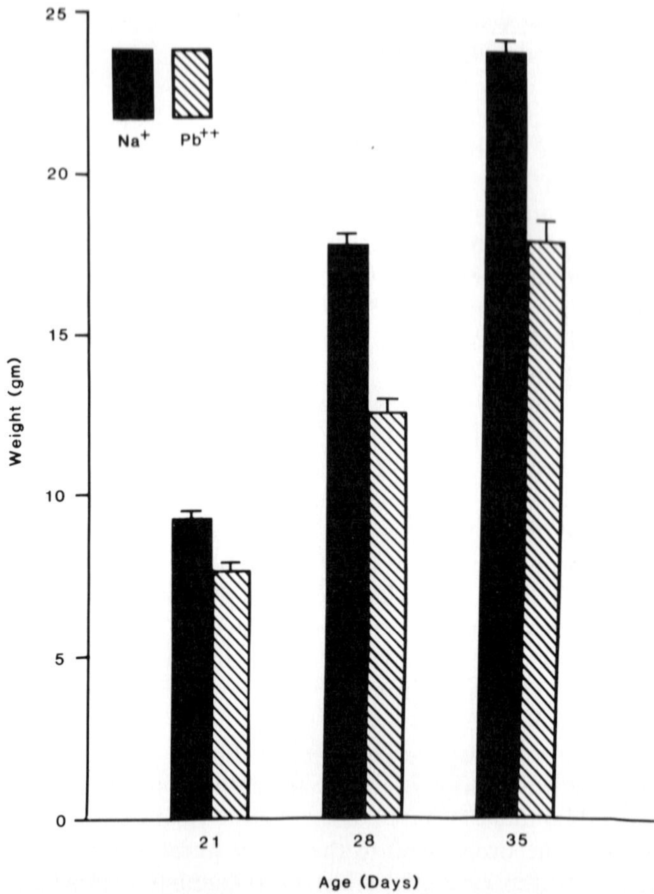

Figure 8.1 Weights of groups of mice exposed to sodium or lead from birth and assessed at 21, 28 and 35 days of age. The bars represent the mean ± SEM. At each time the animals exposed to lead were significantly smaller ($p<0.01$) than were the animals exposed to sodium

Dopamine caused a different pattern of results (Figure 8.2). The lower dose caused the group of animals that had been exposed to sodium to exhibit less locomotor activity than did the animals that had been exposed to saline. This dose also caused the animals that had been exposed to lead to have increased, but not significantly, locomotor activity relative to the animals exposed to sodium. At the larger dose of dopamine the animals exposed to sodium were slightly less active than the saline controls and the group exposed to lead was significantly less active than the group exposed to sodium.

In none of the groups was there a significant change in locomotor activity caused by the drugs when compared with the animals given saline. Thus, neither of the drugs would be expected to be of benefit in children who were hyperactive and thought to have MBD.

The effects of the drugs on the permeability of the BBB to radioactive pertechnetate are illustrated in Figure 8.3. The permeability was significantly decreased in the animals exposed to lead and given saline—i.e. the control group. This relationship was also true for all of the groups given phenylephrine or dopamine, except for the larger dose of dopamine. At this dose the permeability was decreased significantly in the animals exposed to sodium relative to the control group given saline. Phenylephrine (0.125 mg/kg) and dopamine (2 mg/kg) in animals exposed to sodium both caused a significant increase in the permeability of the BBB relative to the control group given saline. This shows that lead can cause changes in the function of the BBB.

DISCUSSION

The finding that exposure to lead resulted in a slower gain in weight agrees with earlier reports (Domer and Llera, 1978; Domer and Wolf, 1979, 1980; Schmidt and Czech, 1977). It is not clear whether this is due only to a decreased intake of nourishment or whether there is an associated change in the rate of development and maturation of the central nervous system (Kalverboer, 1978; Loch *et al.*, 1978). Nevertheless this is a consistent finding.

The fact that neither phenylephrine nor dopamine caused a significant change in the locomotor activity suggests that neither drug would be of benefit in the child with hyperactivity and MBD. This laboratory has previously reported that drugs known to be beneficial in children who are hyperactive—viz. amphetamine, methylphenidate, magnesium pemoline or caffeine—do alter the locomotor activity of mice exposed to lead (Domer and Llera, 1978; Domer and Wolf, 1979, 1980). Thus, this predictor would indicate little likelihood of either drug being of value in treating hyperactive children. As both drugs act on receptors involved in mediating the effects of the sympathetic nervous system (i.e. alpha$_1$ and dopaminergic receptors), it can be inferred either that those receptors are not involved in the beneficial effects caused by amphetamine and the other drugs or that sufficient concentrations of the drugs did not reach the sites where the activity is initiated.

The permeability of the BBB to technetium was decreased by lead in the present experiments. This is in contrast to the previous findings (Domer and Llera, 1978; Domer and Wolf, 1979) and is most likely due to the fact that in the present experiments the blood was removed from the cerebral vessels with saline, whereas in the earlier experiments this was not the case. As with locomotor activity, there was not a significant change in the

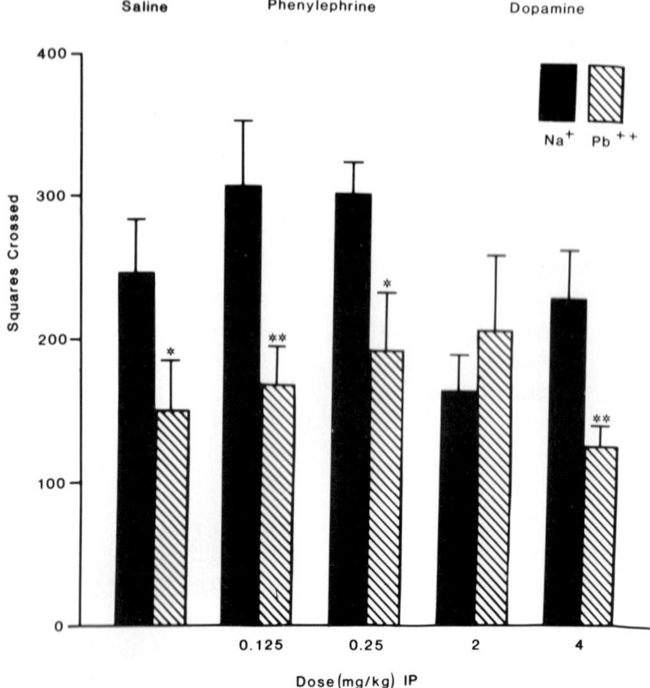

Figure 8.2 The locomotor activity of mice given saline, phenylephrine or dopamine. Groups of ten mice were chronically exposed to either sodium or lead. The significance for each dose of drug given to animals exposed to sodium versus those exposed to lead is indicated by * = <0.05 or ** = <0.01

permeability of the BBB caused by either phenylephrine or dopamine. This would suggest that neither would be expected to be useful in the therapy of the MBD syndrome.

CONCLUSIONS

The MBD syndrome occurs in 5–10% of children under the age of 12. It is characterized by hyperactivity, a short attention span and impulsiveness, which contribute to learning disabilities and social problems. Possible factors contributing to the syndrome include perinatal hypoxia, head injury, malnutrition or exposure to lead. Drugs used to treat MBD often involve activation of receptors activated by the sympathetic nervous system. The present experiments evaluated the potential of phenylephrine and dopamine, two drugs that act on some adrenergic receptors, to modify the changes in mice caused by exposure to lead from birth. The lead caused the young mice to be smaller and less active than other animals exposed to

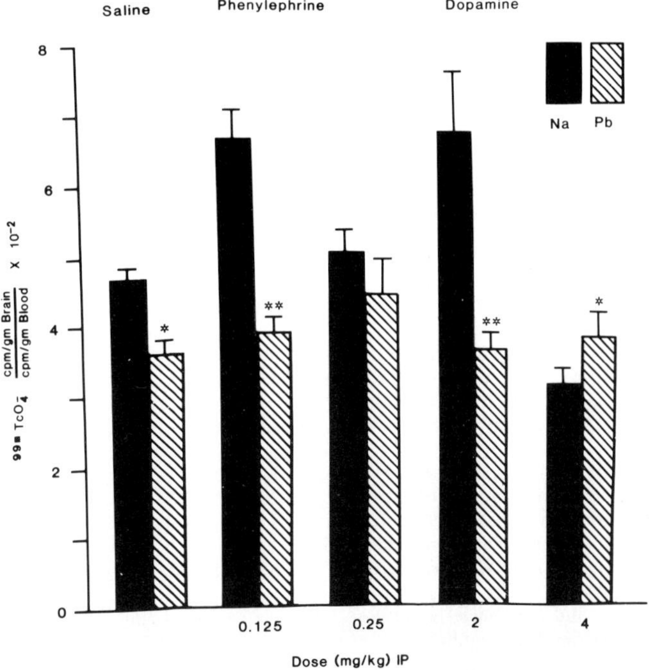

Figure 8.3 The permeability of the blood–brain barrier, as estimated by the ratio of radioactive pertechnetate in brain relative to blood, of mice exposed to sodium or lead from birth. Groups of ten animals each were used. Comparison of the values for each dose of drug of animals exposed to sodium versus lead is indicated by * = <0.05 or ** = <0.01

sodium. Neither drug caused a change in the permeability of the BBB as measured with radioactive pertechnetate. It was concluded that neither would be of benefit in the treatment of the MBD syndrome.

ACKNOWLEDGEMENT

The author would like to thank Mr William Sargent and Mr Eugene Maulet for assistance in these experiments.

REFERENCES

Blackman, S. S. Jr. (1937). The lesions of lead encephalitis in children. *Bull. Johns Hopkins Hosp.*, **6**, 1–61

Bradbury, M. W., Deane, R. and Park, S. (1987). ^{203}Pb uptake into different regions of rat brain, perfused *in situ*. *J. Physiol. (London)*, **390**, 91P

Davson, H., Welch, K. and Segal, M. B. (1987). *The Physiology and Pathophysiology of the Cerebrospinal Fluid*. Churchill Livingstone, New York

Domer, F. R. and Llera, J.-C. (1978). Blood–brain barrier permeability changes caused by lead exposure and amphetamine in mice. *Res. Comm. Psychol. Behav.*, **3**, 101–108

Domer, F. R. and Wolf, C. L. (1979). Drugs, lead and the blood–brain barrier. *Res. Comm. Psychol. Psychiat. Behav.*, **4**, 135–148

Domer, F. R. and Wolf, C. L. (1980). Effects of lead on movement of albumin into brain. *Res. Comm. Chem. Pathol. Pharmacol.*, **29**, 381–384

Hertz, M. M., Bolwig, T. G., Grandjean, P. and Westergaard, E. (1981). Lead poisoning and the blood–brain barrier. *Acta Neurol. Scand.*, **63**, 286–296

Johnson, C. E. (1980). Permeability and vascularity of the developing brain: cerebellum vs. cerebral cortex. *Brain Res.*, **190**, 3–16

Kalverboer, A. F. (1978). MBD: discussion of the concept. *Adv. Biol. Psychiat.*, **1**, 5–17

Kolber, A. R., Krigman, M. R. and Morell, P. (1980). The effect of *in vitro* and *in vivo* lead intoxication on monosaccharide transport in isolated rat brain microvessels. *Brain Res.*, **192**, 513–521

Lefauconnier, J. M., Lavielle, E., Terrien, N., Bernard, G. and Fournier, E. (1980). Effects of various lead doses on some cerebral capillary functions in the suckling rat. *Toxicol. Appl. Pharmacol.*, **55**, 467–476

Lin-Fu, J. (1973). Vulnerability of children to lead exposure and toxicity. *New Engl. J. Med.*, **288**, 1229–1233, 1289–1293

Loch, R. K., Rafales, L. S., Michaelson, I. A. and Bornschein, R. L. (1978). The role of undernutrition in animal models of hyperactivity. *Life Sci.*, **22**, 1963–1970

Lorenzo, A. V. and Gewirtz, M. (1977). Inhibition of [^{14}C] tryptophan transport into brain of lead exposed neonatal rabbits. *Brain Res.*, **132**, 386–392

Mather, K. (1947). *Statistical Analysis in Biology*, 2nd edn. Interscience, New York, pp. 42–44

Michaelson, I. A. and Bradbury, M. (1982). Effect of early inorganic lead exposure on rat blood–brain barrier permeability to tyrosine or choline. *Biochem. Pharmacol.*, **31**, 1881–1885

Miller, E., Sapienza, P., Michel, T. C., Olivito, V. L., Earl, F. L. and Van Loon, E. J. (1979). Some aspects of neurotoxic effects of lead in neonate beagle dogs. *Fed. Proc.*, **34**, 267

Millichap, J. G. (1982). The hyperactive child. *Triangle*, **21**, 59–63

Mitchell, W., Kim, C., O'Tuama, L. and Pritchard, J. (1980). Blood–brain barrier and maturation affects susceptibility to lead. *Neurology*, **30**, 353

Pentschew, A. (1965). Morphology and morphogenesis of lead encephalopathy. *Acta Neuropathol.*, **5**, 133–160

Pentschew, A. and Garro, F. (1966). Lead encephalo-myelopathy of the suckling rat and its implications on the porphyrinopathic nervous diseases. With special reference to permeability disorders of the nervous system's capillaries. *Acta Neuropathol.*, **6**, 266–278

Press, M. F. (1977). Lead encephalopathy in neonatal Long–Evans rats: morphological studies. *J. Neuropathol.*, **36**, 169–193

Pyck, K. and Baines, P. (1978). The influence of drugs on minimal brain dysfunction. *Adv. Biol. Psychiat.*, **1**, 68–83

Raimondo, A. J., Beckman, F. and Evans, J. P. (1968). Fine structural changes in human lead encephalopathy. *J. Neuropathol.*, **27**, 154

Sandstrom, R., Muntzing, K., Kalimo, H. and Sourander, P. (1985). Changes in the integrity of the blood–brain barrier in suckling rats with low dose lead encephalopathy. *Acta Neuropathol.*, **68**, 1–9

Schmidt, J. C. and Czech, D. A. (1977). Effect of tetraethyl lead and restricted food intake on locomotor activity in the rat. *Pharmacol. Biochem. Behav.*, **7**, 489–492

Shaffer, D. (1978). Longitudinal research and the minimal brain damage syndrome. *Adv. Biol. Psychiat.*, **1**, 18–34

Silbergeld, E. K. and Goldberg, A. M. (1973). A lead-induced behavioral disorder. *Life Sci.*, **13**, 1275–1283

Smith, J. F., McLaurin, R. L., Nichols, J. B. and Asbury, A. (1960). Studies in cerebral oedema and cerebral swelling I. The changes in lead encephalopathy in children compared with those in alkyl tin poisoning in animals. *Brain*, **83**, 411–424

Suckling, A. J., Rumsby, M. G. and Bradbury, M. W. B. (Eds) (1986). *The Blood–Brain Barrier in Health and Disease*. Ellis Horwood, Chichester

Thomas, J. A., Dallenbach, F. D. and Thomas, M. (1973). The distribution of radioactive lead (^{210}Pb) in the cerebellum of developing rats. *J. Pathol.*, **109**, 45–50

Toews, A. D., Kolber, A., Hayward, J., Krigman, M. A. and Morell, P. (1978). Experimental lead encephalopathy in the suckling rat: concentration of lead in cellular fractions enriched in brain capillaries. *Brain Res.*, **147**, 131–138

Turnbull, J. and Brodeur, J. (1984). Influence d'un retard de croissance sur le permeabilité de la barriére hémato-encephalique chez le rat nouveau-né exposé au plomb. *Can. J. Physiol. Pharmacol.*, **62**, 142–145

Zenick, H., Lasley, S. M., Greenland, R., Caruso, V., Succop, P., Price, D. and Michaelson, I. A. (1982). Regional brain distribution of d-amphetamine in lead-exposed rats. *Toxicol. Appl. Pharmacol.*, **64**, 52–63

9

Mechanisms Regulating Peptide Levels in the Cerebrospinal Fluid

David J. Begley and Daniel G. Chain

Presented at a symposium held to mark the occasion of Hugh Davson's 80th birthday at The Sherrington School of Physiology, United Medical and Dental Schools of Guy's and St. Thomas's Hospitals.

INTRODUCTION

A large number of regulatory peptides are now recognized as naturally occurring in the cerebrospinal fluid (CSF) and central nervous system (CNS) (Tables 9.1 and 9.2).

Table 9.1 Neuropeptides occurring in brain tissue

ACTH	GnRh
Angiotensin II	α-MSH
Bombesin	Motilin
Bradykinin	Neurokinin A
Calcitonin	Neurokinin B
Calcitonin gene-related peptide	Neuropeptide Y
Carnosine	Neuropeptide P
Cholecystokinin	Neurotensin
Corticotropin releasing factor (CRF)	Oxytocin
Dynorphin	Peptide HI
β-Endorphin	Prolactin
[Met^5]-Enkephalin	Proctolin
[Leu^5]-Enkephalin	Secretin
Galanin	Somatostatin
Gastrin	Vasopressin
Glucagon	TRH
Growth hormone (somatotropin)	Substance P
Growth hormone releasing hormone	Insulin
Vasoactive intestinal polypeptide (VIP)	Kyotorphin
Endothelin 1	Lipotropin
Endothelin 3	

Adapted from Iversen (1987).

Table 9.2 Neuropeptides present in the cerebrospinal fluid

Neuropeptide	*Other locations*
GnRH	Gonad
Somatostatin	Pancreas
[Met^5]-Enkephalin	Intestine, adrenal medulla
[Leu^5]-Enkephalin	Intestine, adrenal medulla
β-Endorphin	Adrenal medulla
Dynorphin	Adrenal medulla
Insulin	Pancreas
TRH	
CRH	
GHRH	
Gastrin	Stomach
CCK	Intestine
VIP	Intestine
Angiotensin II	Lung (conversion)
Substance P	
DSIP	Gut
Neurotensin	
Oxytocin	
Vasopressin	
Bombesin	Gut
Carnosine	
Adenohypophyseal hormones	

Adapted from Jackson (1980).

These regulatory peptides may gain entry to the CSF by production locally by central nervous structures, followed by release into the CSF; or by translocation from plasma across the blood–brain barrier (BBB) or the blood–CSF barrier (BCSFB). Local production would also include production by cells of the central nervous system as a classical neurotransmitter with escape into CSF or as a direct paracrine secretion into the CSF.

Characteristic actions of these peptides can be seen either after intraventriculocerebral injection or after peripheral administration. Often the effect is greatest after intracerebroventricular administration, suggesting that access to the targets for the peptides from blood is restricted and regulated.

The presence of a selective blood–brain barrier to regulatory peptides has two important implications. First, if the barrier is selectively permeable, it will allow only certain blood-borne peptides to have access to the CNS, enabling them to exert their effects. Second, if the barrier is selectively impermeable to specific peptides, it enables these peptides to exert separate and independent actions within and outside of the CNS.

Thus, in some cases the brain is able to regulate peptide concentrations in CSF separately and independently from that of the other extracellular fluids and thus enabling the central nervous system to maintain its own hormonal and regulatory milieu.

Whatever the origins of the regulatory peptides in CSF, a number of mechanisms will serve to regulate their levels in this fluid compartment. These will fall into three general areas which are discussed with examples. These mechanisms are: (1) turnover of CSF volume by secretion bulk flow and drainage; (2) the enzymatic inactivation/activation of peptides by hydrolysis; and (3) specific transport mechanisms.

The examples that are used to illustrate the general principles of these mechanisms and their relative importance are drawn from work performed in conjunction with Hugh Davson between 1980 and 1984, using a ventriculocisternal perfusion technique to study the transport of glycyl-L-leucine, thyrotropin releasing hormone (TRH; pyroglutamyl-histidyl-prolinamide) and leucine enkephalin (tyrosyl-glycyl-glycyl-phenylalanyl-leucine) and other solutes and analogues, from the cerebrospinal fluid of the rabbit. The perfusion technique used has been fully described by Davson *et al.* (1982).

Briefly, artificial CSF containing radiolabelled tracer was introduced into both lateral ventricles of the rabbit at a constant rate of approximately 67 μl/min and the effluent perfusate was collected at the cisterna magna under a negative pressure of 5–7 cm of water. Under these conditions the perfused area is limited to the lateral, third and fourth ventricles and the cisternal space, with a perfused volume of about 1.5 ml.

A large molecular weight dextran (2 million) which is not subject to any transport processes is included in the perfusate and this enables the dilution of the perfused fluid by newly formed CSF to be estimated. By this method the rate of CSF production in the perfused area of the rabbit is 10.5 ± 0.7 ml/min. If in the intact animal the CSF volume is to remain constant, this rate of production will equal the rate of turnover of fluid and will approximate to the rate of drainage of CSF by bulk flow. This continuous turnover of CSF by bulk flow produces a diffusive sink into the CSF for solutes entering from blood (Davson *et al.*, 1963). If there were no bulk flow, a solute with a finite permeability at the blood–brain interface would with time reach an equilibrium CSF level equal to that of plasma. The sink effect prevents this happening and maintains the equilibrium brain ISF solute level lower than the corresponding plasma level. Transport mechanisms out of CSF for solutes are superimposed upon this phenomenon.

The calculated turnover time of the perfused space by CSF secretion and bulk flow is approximately 143 min, under the conditions of ventriculo-cisternal perfusion. As the volume of the ventricles in the intact animal may be somewhat larger than in the perfused animal, approximating 1.8 ml, this calculated turnover time may be closer to 172 min under normal circumstances.

During ventriculocisternal perfusion, tracer uptake into the choroid plexus and transport into blood occurs. Also, there is penetration into brain extracellular space (interstitial fluid; ISF) and uptake into brain cells. Transport from brain ISF and into blood across the brain capillary endothelium is also possible.

The rate of loss of a peptide or amino acid from the perfusion fluid may be expressed as a clearance, which is defined as the volume, in μl, of perfusion fluid of average concentration, cleared of the tracer per minute. The calculation was made according to the method of Bradbury and Davson (1964), using the following formula:

$$\text{clearance} = \frac{F_{in}(C_{in} - 1/R_d \times C_{out})}{(C_{in} + C_{out})/2} \; \mu l/min$$

where C_{in} is the radioactive concentration of tracer per unit volume of perfusion fluid entering the ventricles, F_{in} is the rate of perfusion, in $\mu l/min$, and R_d is the steady-state value of C_{in}/C_{out} for the 2 million molecular weight blue dextran.

Also a steady-state ratio for labelled tracer can be calculated from the last four samples of perfusion fluid collected at the end of an experiment when the system had reached a steady state as:

$$\text{steady-state ratio, } R_{csf} = (C_{out}/C_{in}) \cdot 100$$

Clearances for the three peptides investigated were: glycyl-L-leucine $54.2 \pm 2.5 \; \mu l/min$ (Begley *et al.*, 1980); leucine enkephalin $20.0 \pm 1.7 \; \mu l/min$ (Begley and Chain, 1981); and thyrotropin releasing hormone (TRH) $8.4 \pm 1.2 \; \mu l/min$ (Begley and Chain, 1982).

CSF TURNOVER

Figure 9.1 shows a typical experiment using blue dextran (MW 2 million), [^{3}H]-sucrose and [^{14}C]-α-aminoisobutyric acid (AIB) as tracers in the perfusate and plotting R_{csf} against time. The dextran, as it is an inert polar molecule, is not subject to any transport processes and comes into equilibrium quickly, being simply diluted by the production of new CSF at the choroid plexuses. Sucrose finds a larger space in the brain, owing to its ability to penetrate into brain extracellular space but typically does not enter cells or interact with any transport mechanisms. Thus, a small peptide, not subject to any transport processes and cleared from CSF by bulk flow, might be expected to give an R_{csf} similar to that of sucrose. The [^{14}C]-AIB is transported rapidly out of the CSF by choroidal transport into blood and by uptake into brain cells, and thus at equilibrium its R_{csf} is much lower than that of sucrose.

During perfusion the perfusate may be exchanged for one which does not contain any labelled tracer, thus enabling the washout of sucrose and AIB to be observed (Figure 9.1). If perfusion is continued for up to 140 min, some 90% of the sucrose that had left the perfusate during the initial part of the experiment can be recovered in the outflow. Thus, the sucrose has remained in the brain interstitial fluid (ISF) probably in a location close to the ventricle and can be washed out rapidly. As the washout experiment

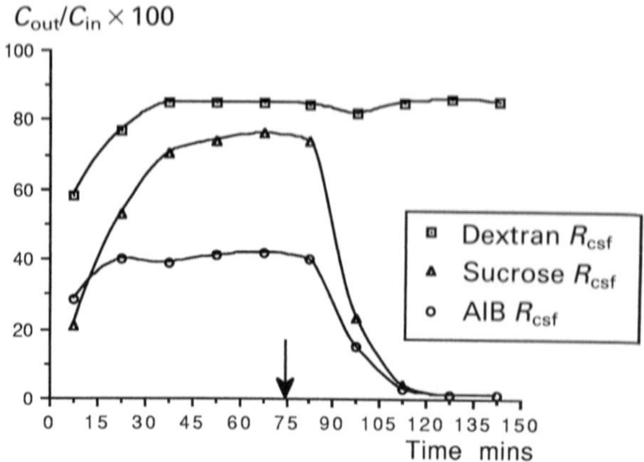

Figure 9.1 A ventriculocisternal perfusion employing blue dextran, [³H]-sucrose and [¹⁴C]-α-aminoisobutyric acid (AIB), as tracers. The R_{csf} (inflow/outflow ratio) is plotted as the abscissa and time as the ordinate. At the arrow 75 min after the start of perfusion the inflowing perfusate is exchanged for one containing blue dextran but no radiolabelled tracer, thus enabling the washout of sucrose and AIB to be followed

continues, the rate of recovery of sucrose diminishes but clearly recovery is continuing beyond 140 min and more of the sucrose would be recoverable with time. This continued but small recovery of sucrose at longer times may indicate the existence of two pools of sucrose in the brain after ventriculocisternal perfusion, one rapidly accessible and adjacent to the ventricles which washes out rapidly and a less accessible pool which washes out more slowly. The extent of penetration of tracers such as sucrose into brain tissue during ventriculocisternal perfusion may be limited by a small but continuous production of ISF by the brain substance tending to produce a bulk flow of ISF from the interstitial spaces into the ventricles across the ependyma, as suggested by Cserr *et al.* (1981).

When the non-metabolizable amino acid AIB is used as tracer, only 32% of the radioactivity lost during the perfusion is recoverable and here tracer can be shown to have left the perfused area and entered blood and is accumulated by cells into a compartment that cannot be washed out. Similar considerations will also apply to the other non-metabolizable amino acid, cycloleucine.

Table 9.3 shows washout data for a number of tracers. Only in the cases of sucrose, AIB and cycloleucine, where the tracer is non-metabolizable under the conditions of the experiment, can one be sure that intact tracer bearing the radiolabel is being recovered. However, the figures are of some interest and it is surprising to note that a considerable amount of leucine-enkephalin radioactivity can be recovered, more than, for instance, with tyrosine, which is the labelled residue in the peptide. The comparison of a ³H- and ¹⁴C-labelled tracer under these circumstances may

Table 9.3 Tracer washout

[^{3}H]-Sucrose	90.0%
[^{3}H]-Leu-enkephalin	43.0%
[^{14}C]-Tyrosine	37.0%
[^{14}C]-AIB	32.0%
[^{14}C]-Cycloleucine	20.0%
[^{14}C]-Glycine	16.5%
[^{3}H]-Leucine	15.0%

not be justified, as, once the tracer is internalized within cells, the resulting metabolic products bearing the two isotopic labels may be very different. In this context it is also interesting to note that the percentage washout and recovery for the two non-metabolizable amino acids AIB and cycloleucine are greater than for the two naturally occurring amino acids glycine and leucine.

ENZYMIC CONVERSION OF PEPTIDES IN CSF

A variety of peptidases have been demonstrated in central nervous tissue which are capable of hydrolysing and terminating the actions of neuropeptides. Some of these enzymes are membrane- and tissue-bound and others are soluble and exist in the highest concentrations in the cytosol (Turner, 1987). If some of these enzymes exist in solution in the CSF, they could exert a scavenging action on susceptible peptides, in addition to mechanisms transporting peptides out of the cerebrospinal fluid.

These peptidases can be classified into three groups.

(1) A membrane-bound neutral metalloendopeptidase generally referred to as endopeptidase 24:11 and formally called enkephalinase A. Endopeptidase 24:11 includes as substrates, [leu^5]-enkephalin, [leu^5]-enkephalinamide, dynorphin(1–9)-peptide, dynorphin(1–13)-peptide, α-neo-endorphin, b-neo-endorphin, [met^5]-enkephalin, [met^5]-enkephalin-arg^6, [met^5]-enkephalin-arg^6-phe^7, β-lipotropin(61–69)-peptide and γ-endorphin (Turner, 1987).

(2) A further group of aminopeptidases, including an extrinsic membrane-associated aminopetidase which is puromycin-sensitive termed aminopeptidase MII, a soluble form of which also occurs in the cytosol (Dyer *et al.*, 1990), and an integral membrane aminopeptidase, aminopeptidase M (Turner, 1987), have been described.

(3) Additionally there is also angiotensin converting enzyme (ACE), an extrinsic membrane-associated peptide of neuroepithelial cells, including the luminal surface of the brain capillary endothelium and the microvilli of the choroid plexus (Rush and Hersh, 1982). ACE is also found associated with the membranes of neurons but not glia (Skidgel *et al.*, 1987) and is

also present in the cerebrospinal fluid as a soluble enzyme (Schelling *et al.*, 1980). Combined with its best-known action of converting angiotensin I and angiotensin II, ACE is also catholic in its choice of substrates and will hydrolyse as alternative substrates to A I, bradykinin, des-arg^8-bradykinin, [leu^5]- and [met^5]-enkephalin, [met^5]-enkephalin-arg^6-phe^7, [met^5]-enkephalin-arg^6-gly^7-leu^8, β-neo-endorphin, dynorphin(1–8), dynorphin(1–6), neurotensin, substance P and LHRH (Skidgel *et al.*, 1987).

In the case of thyrotropin releasing hormone (TRH), enzymatic hydrolysis is rapid in plasma (Redding and Schally, 1972), producing the free acid (TRH-OH) and His-pro-NH$_2$. Hydrolysis also may take place at the blood–brain interfaces of the cerebral capillaries and the choroid plexus (Zloković *et al.*, 1985, 1988). Both brain tissue and brain homogenates contain deamidase enzymes that will convert TRH to the free acid (Hersh and McKelvy, 1979; Matsui *et al.*, 1979) and also to histidine and prolinamide (Matsui *et al.*, 1979). The brain enzymes include a pyroglutamate aminopeptidase I, a proline endopeptidase and a dipeptidyl peptidase in the cytoplasmic fractions of brain homogenates; an iminopeptidase in the particular fraction; and a pyroglutamate aminopeptidase II and dipeptidyl aminopeptidase II in association with synaptosomal membranes (O'Cuinn *et al.*, 1990).

Although it is generally accepted that the neurotransmitter–neuromodulatory actions of enkephalins and probably other neuropeptides are terminated by enzymic degradation (Schwartz *et al.*, 1981; Hersh, 1982, 1985; Dyer *et al.*, 1990), this activity is probably most important in brain interstitial fluid, and the more general questions of bulk flow versus hydrolysis and transport mechanisms take on a greater importance in the cerebrospinal fluid. Indeed, when it comes to the question of peptide entry to the CNS, peptides that may have been transported across the brain capillary endothelium by a process of transcytosis will be most subject to the enzymic influences in brain ISF and on the cell membranes of neurons and glia, and peptides that may have gained access to the central nervous system across the choroid plexus will be more subject to the influences of CSF factors.

With regard, then, to the three peptides compared in the present chapter—namely, glycyl-L-leucine, TRH and leucine-enkephalin—the enzymic activity of CSF was investigated by incubating the labelled peptide for 1 h with cerebrospinal fluid at 37 °C and examining the incubation medium for possible breakdown products by either thin layer chromatography, thin layer ion exchange chromatography or thin layer electrophoresis. The results are shown in Figures 9.2, 9.3 and 9.4. Of the three peptides, only leucine-enkephalin shows evidence of breakdown products after these experiments. Figure 9.2 shows the effect of incubation of [^{14}C]-glycyl-L-leucine with cerebrospinal fluid. After 1 h of incubation, all of the

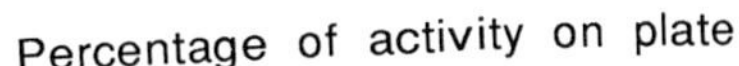

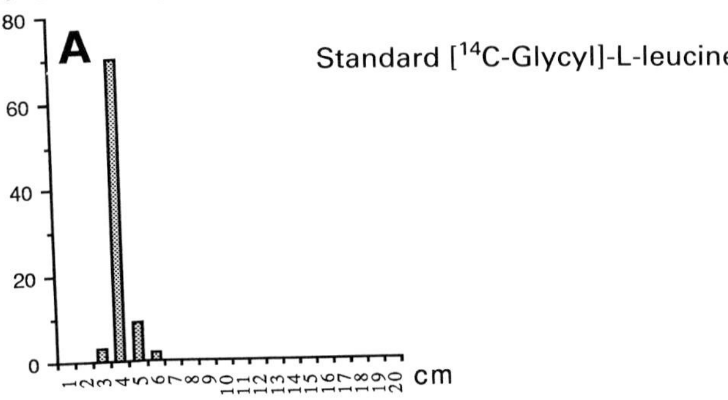

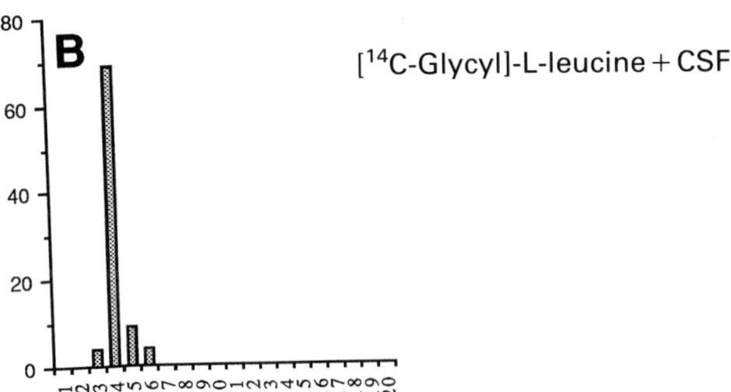

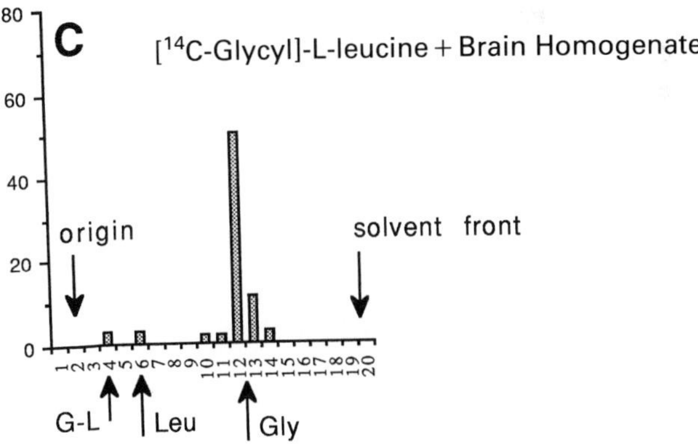

Figure 9.2 Incubation of [^{14}C]-glycyl-L-leucine with rabbit cerebrospinal fluid. Thin layer ion exchange chromatography was performed on Ionex-25 SA-Na, 5×20 precoated plates, Machery-Nagel, using citrate buffer pH 3.3. (A) Shows the separation of a [^{14}C]-glycyl-L-leucine standard. (B) Illustrates the same but after 1 h incubation with 0.5 ml CSF diluted with 0.5 ml HEPES-buffered saline, 15 mM, pH 7.4, indicating no breakdown of the dipeptide. (C) After 1 h incubation with 100 mg of brain homogenate dispersed in 1 ml of HEPES-buffered saline. Here most of the dipeptide has been hydrolysed and the radioactivity appears at the glycine R_f. All incubations at 37 °C

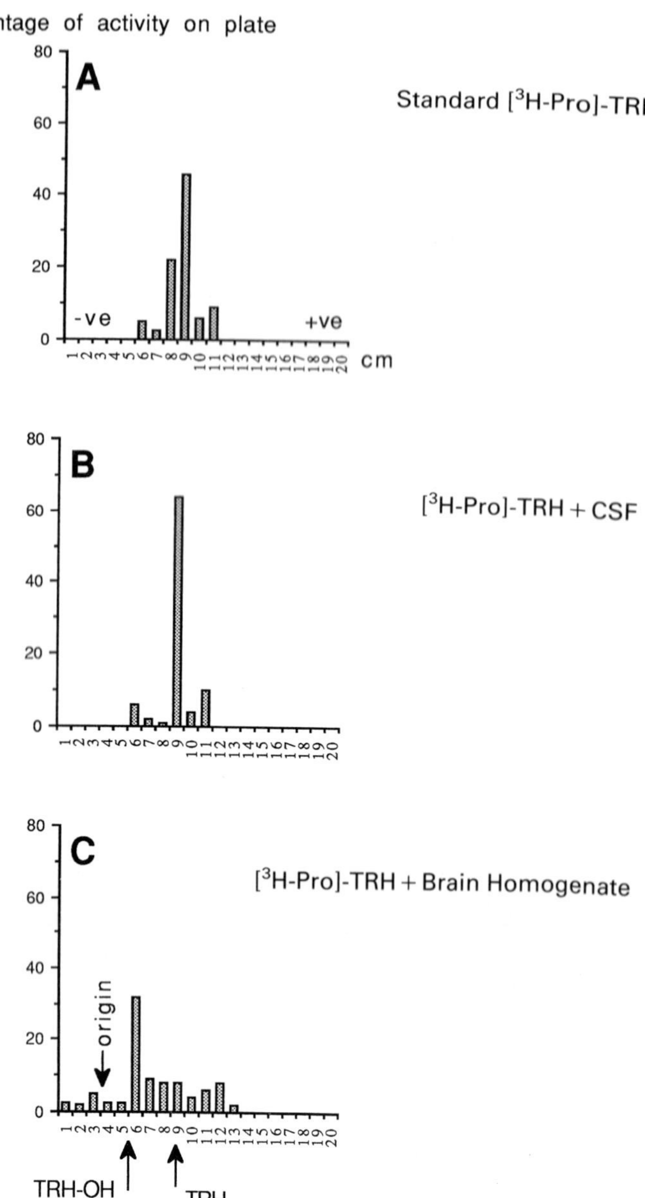

Figure 9.3 Incubation of [³H-prolinamide]-TRH with rabbit cerebrospinal fluid. Electrophoresis was performed on cellulose acetate electrophoresis strips, 5×20, Oxoid Ltd. Electrophoresis was performed in pyridine/acetic acid buffer pH 3.5 at 15 V/cm, for 2 h. (A) Shows the separation of a [³H]-TRH standard. (B) The same but after 1 h incubation with CSF. (C) After incubation with brain homogenate, showing breakdown of the TRH and generation of considerable quantities of the free acid TRH-OH and other labelled products. Details of incubation procedure as in Figure 9.2

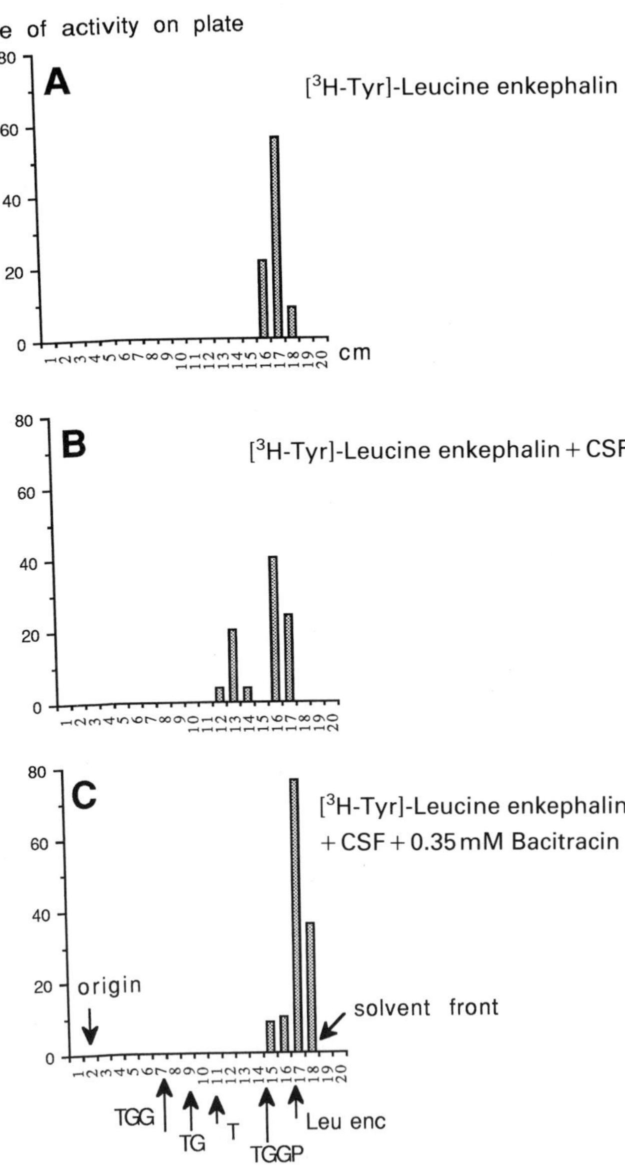

Figure 9.4 Incubation of [³H-Tyr]-leucine-enkephalin with rabbit cerebrospinal fluid. Thin layer chromatography was performed on Polygram Sil-G, 5 × 20, precoated plates, Machery-Nagel, with isopropanol:ethyl acetate:5% acetic acid (2:2:1) as the mobile phase. (A) Shows the separation of a [³H-Tyr]-leucine standard. (B) The same but after 1 h incubation with CSF. Note that a large proportion of the [¹⁴C] activity is no longer at the leucine-enkephalin R_f but lies closest to the TGGP R_f. (C) The same but with CSF containing 0.35 mM bacitracin, demonstrating that this inhibitor of peptidase activity prevents the breakdown of leucine-enkephalin under the conditions of the experiment. T = tyrosine; TG = tyrosyl, glycine; TGG = tyrosyl, glycyl, glycine; TGGP = tyrosyl, glycyl, glycyl, glycine. Details of incubation procedure as in Figure 9.2

radioactivity remained at the R_f for glycyl-L-leucine with no radioactivity appearing at the glycine R_f. After a similar incubation for 1 h with brain homogenate, virtually all of the radioactivity is at the glycine R_f and this hydrolysis did not appear to be sensitive to the inclusion of bacitracin 0.35 mM in the incubation medium.

Figure 9.3 shows the result of a similar experiment with [^{3}H]-TRH, using an electrophoretic separation which is more effective at separating TRH and its free acid TRH-OH. After 1 h incubation with CSF, there is no evidence of conversion to TRH-OH and the radioactivity is still localized at the TRH R_f. When TRH is incubated with brain homogenate, however, there is a considerable amount of radioactivity present at the TRH-OH R_f and at other locations on the thin layer plate. As TRH is labelled at the prolinamide residue, some of this activity may represent free proline or prolinamide and also formation of the diketopiperazine cyclo-histidyl-proline. The R_fs for these substances were not determined. The cyclization of the N-terminus of TRH to form pyroglutamic acid and the amidation of the C-terminal proline to form prolinamide may confer a degree of protection against enzymic breakdown of TRH to the peptidases present in CSF (Begley and Chain, 1982), whereas little protection against hydrolysis is afforded to TRH by these modifications to enzymes present in plasma and at the blood–brain interface (Zloković *et al.*, 1985, 1988).

This blocking of the N and C termini protects the TRH molecule from attack by exopeptidases. Also, deamidation to form the free acid does not occur in CSF. Cyclization of the N terminus also has important implications regarding interaction with the carriers of transport systems. It is interesting that a number of other neuropeptides which occur in CSF show similar modifications. Both bombesin and GnRH have both a cyclized pyroglutamic N terminus and an amidated C terminus, whereas neurotensin has a pyroglutamic acid N-terminal residue and substance P, ACTH, angiotensin II, oxytocin, arginine-vasopressin, vasoactive intestinal polypeptide (VIP) and CCK 8 all have amidated C termini.

Figure 9.4 shows the results of the experiments with [^{3}H-Tyr]-leucine-enkephalin. After a 1 h incubation with CSF, a considerable amount of radioactivity remains at the leucine-enkephalin R_f but there is a generation of radioactivity at a position closely approximating to the R_f for tyrosyl-glycyl-glycyl-phenylalanine, (des-[leu^5]-leucine-enkephalin), presumably generated by a carboxypeptidase activity. There is no evidence for the generation of free tyrosine, tyrosyl-glycine or tyrosyl-glycyl-glycine, which are the other possible labelled products (Begley and Chain, 1988). Hence, for the peptides under discussion only leucine-enkephalin appears to exhibit moderate hydrolysis in CSF; this hydrolysis is abolished by the addition of 0.35 mM bacitracin into the incubation medium.

With regard to the situation in a ventriculocisternal perfusion experiment, during perfusion the endogenous CSF is of course flushed out of the perfused compartments and thus the enzyme activity from this source is greatly reduced, with enzyme possibly only being present in the newly secreted nascent CSF. Experiments similar to those described above designed to detect products of hydrolysis in the outflowing CSF failed to detect radioactivity other than in the intact tracer. Addition of bacitracin to the inflowing perfusate or during a perfusion by switching syringes (Figure 9.5) produced no change, indicating little residual hydrolysis from soluble enzymes or membrane-bound enzymes on ependymal or other cells with which the labelled tracer may come into contact. In contrast, the luminal membranes of the cerebral capillary endothelium possess membrane-bound enzymes capable of hydrolysing TRH but not leucine-enkephalin (Zloković *et al.*, 1987; Segal and Zloković, 1990).

These experiments indicate, therefore, that under the conditions of ventriculocisternal perfusion, for the peptides studied, hydrolysis with the uptake or transport of labelled products is not the rate-limiting step in the clearance of peptide from CSF. Even in the intact animal with leucine-enkephalin the contribution of enzymic activity is likely to be low as more than 70% of enkephalin added to CSF and incubated remains at the enkephalin R_f after 1 h. The peptidases in brain homogenate, although it may be possible to classify them as soluble peptidases, may be largely restricted to the cytosol and cellular internalization of peptide is necessary for their activity. It is important to bear in mind that both peptides and

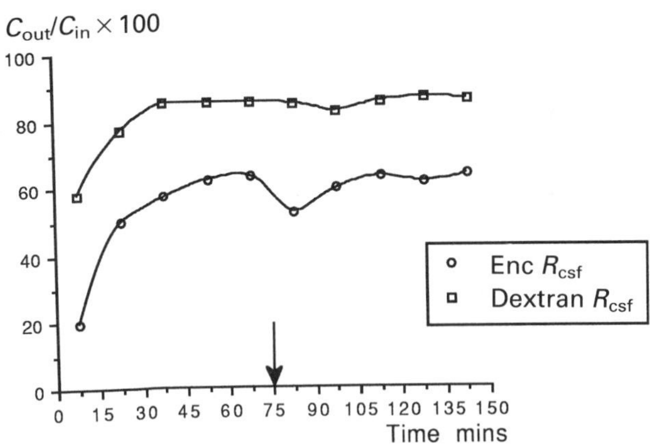

Figure 9.5 A ventriculocisternal perfusion in the rabbit employing blue dextran and [^{3}H-Tyr]-leucine-enkephalin as tracers. The R_{csf} (inflow/outflow ratio) is plotted as the abscissa and time as the ordinate. At the arrow 75 min after the start of perfusion the inflowing perfusate is exchanged for one containing blue dextran and the labelled enkephalin but also 0.35 mM bacitracin. The addition of bacitracin to the perfusate, part way through a perfusion, has no effect on the R_{csf} for leucine-enkephalin, indicating that under the conditions of perfusion, hydrolysis of the peptide is not a rate-limiting step in clearance. The temporary drop in the R_{csf} immediately after the exchange of perfusate is the result of a transient interruption of flow in the perfusate resulting in a rise in the calculated clearance rate until equilibrium is re-established

peptidases are highly compartmentalized in the CNS and that their contact will be highly regulated. Thus, the behaviour of a peptide in CSF and brain ISF will depend on this compartmentalization and access of peptidase to substrate.

SPECIFIC TRANSPORT MECHANISMS

Figure 9.6 shows total clearance (percentage of total tracer lost from the perfused system), together with clearance to brain tissue and to blood, during a 120 min perfusion. Total clearance of tracer is calculated as the difference between the amount perfused into the ventricles, $F_{in}{\cdot}C_{in}{\cdot}t$, and the amount of tracer collected in the outflowing perfusate at the cisterna magna,

$$\sum_{t=0}^{t=120} C_{out}{\cdot}t$$

The quantity in brain is calculated from the brain dpm per unit weight of brain and the whole brain weight after removal from the skull. The clearance to blood is calculated as the difference between the total tracer activity perfused and that recovered in both the outflowing perfusate and brain. In all cases losses are calculated relative to the activity of the inflowing perfusate (arbitrarily set at 100%).

In Figure 9.6(A), when sucrose is employed as the labelled tracer, all of the tracer that has been lost can be recovered from the animal's brain at the end of the experiment. Thus, there is no net removal of sucrose from the system during the experiment and no loss of tracer to blood or other compartments. Sucrose is thus acting typically as an extracellular marker and the clearance over and above that of dextran is the result of a greater degree of penetration into brain extracellular space across the ventricular ependyma. The washout data already presented (Table 9.3) show that this sucrose is easily recovered and is thus remaining extracellular and adjacent to the ventricle. If this observation is true for most tracers during ventriculocisternal perfusion, the implication will be that most tracers which leave the perfused volume do so largely at the choroid plexus or by uptake into adjacent brain tissue and not generally at the abluminal surfaces of the brain capillary endothelia, most of which will be a considerable diffusional distance from the ventricles through brain ISF, and this diffusion may be countercurrent to the flow of brain extracellular fluid.

In Figure 9.6(B), when TRH is employed as tracer, there is a small clearance to blood, again probably largely across the choroid plexus. In total, however, about twice as much tracer, in terms of that introduced into the ventricle, is cleared from CSF during a perfusion than is sucrose. As little is transported to blood, the larger space found by TRH suggests an

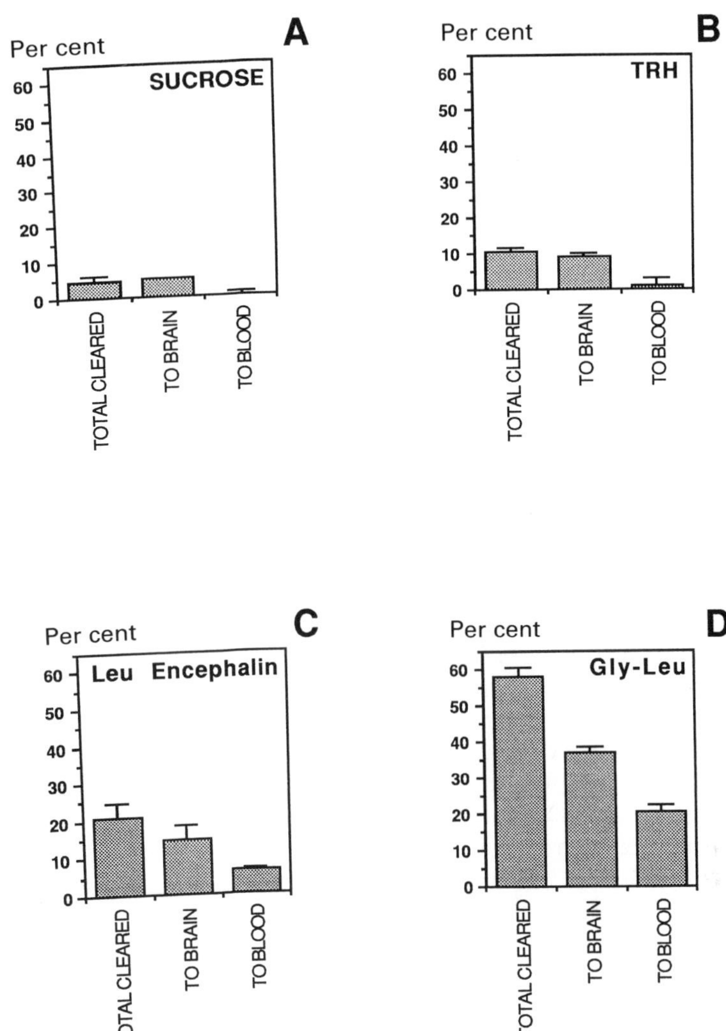

Figure 9.6 Panels A, B, C and D show the losses of radioactively labelled tracer during a 120 min ventriculocisternal perfusion in the rabbit. The total clearance is the total loss of tracer between inflowing and outflowing perfusion fluid during the experiment and its calculation is described in the text. The clearance to brain is the proportion of radioactive label that can be recovered in brain tissue at the end of an experiment. The loss to blood represents tracer that has left the perfused area entirely and is obtained by the difference between the total clearance and the tracer activity recoverable in brain. (A) Sucrose; virtually all of the tracer cleared is recoverable from brain with a zero clearance to blood. (B) TRH; the total amount cleared is larger than that for sucrose. Almost all of the cleared tracer can be recovered from brain but there is a small loss to blood. (C) Leucine-enkephalin; here the amount cleared is considerably greater, with some two-thirds of tracer radioactivity recoverable from brain and the remainder cleared to blood. (D) Glycyl-L-leucine; a considerable amount of the dipeptide is cleared from the perfusate, with again some two-thirds of tracer radioactivity being recoverable from brain, the rest being lost to blood

accumulation by cells, either by internalization or by accumulation onto surface receptors.

Turning to leucine-enkephalin (Figure 9.6C), there is a marked clearance of the tracer during ventriculocisternal perfusion, with the majority of the tracer radioactivity being accumulated by brain, indicating internalization or accumulation onto cellular membrane receptors, together with a substantial clearance to blood.

With the artificial dipeptide glycyl-L-leucine (Figure 9.6D), there is a considerable total clearance from the CSF, with two-thirds of the perfused material being accumulated by brain tissue and the rest being cleared to blood.

ANALYSING TRANSPORT MECHANISMS

A number of distinct transport mechanisms can be shown to contribute to the transport of peptides from CSF. Some of these are clearly related to the amino acid transporters originally described by Christensen (1973); others appear more specific to peptides (Banks *et al.*, 1986; Begley and Chain, 1988; Banks and Kastin, 1990). By introducing combinations of competitive inhibitors into the perfusate, it is possible to model these systems.

It is important to recognize that in the ventriculocisternal perfusion system it is not possible to identify precisely the sites of transport (Cserr, 1971), although, as previously emphasized, these seem to be sited in tissues immediately adjacent to the cerebral ventricles and also at the choroid plexuses. Also, because the interfaces are not homogeneous, it is probably unlikely that single carrier systems occur and that multiple carriers for a single labelled tracer will exist and the carriers may be distributed in different combinations in different parts of the CNS. A hypothetical example of this argument is illustrated in Figure 9.7. Here the primary transporter (4) is the L-system of Christensen but, in addition, transporters (1) to (6) are also present. BCH (2-amino-bicyclo-(2,2,1)heptane-2-carboxylic acid) is the defining substrate for Christensen's L-system. Hence, BCH, leucine and tyrosine will all be transported by system (4) but, in addition, to varied extents leucine is transported by systems (1), (3) and (6) and tyrosine by systems (1), (2) and (5). Hence, all three tracers introduced at 5 mM concentration into the perfusate will substantially inhibit clearance of the others but self-inhibition will always be more effective than cross-inhibition between tracers, as the full spectrum of carriers will be inhibited (Davson *et al.*, 1982).

Thus, a comparative approach of self- and competitive inhibition can be used to construct models for peptide transport.

Figure 9.8 shows the results obtained with inhibition experiments performed with glycyl-L-leucine. In these experiments either the glycine or

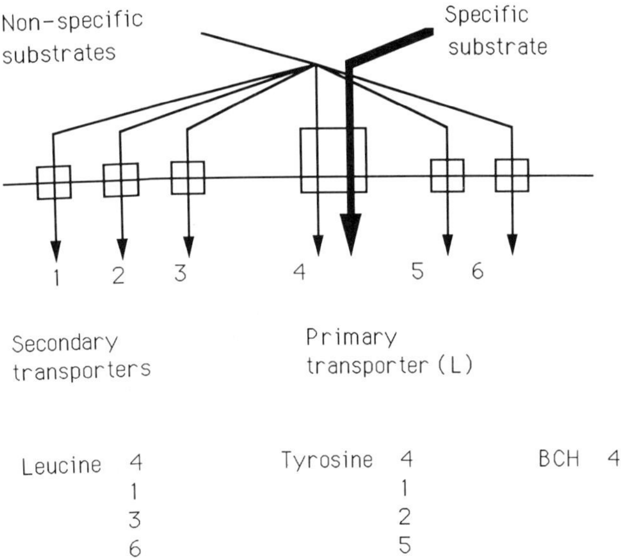

Figure 9.7 Hypothetical scheme illustrating the implications of multiple transporters for amino acids in the central nervous system. Of six postulated transporters, only one represents the L-system of Christensen and thus is specific for BCH (2-amino-bicyclo-(2,2,1)heptane-2-caboxylic acid). Hypothetical spectra of transporter reactivity for leucine and tyrosine are also suggested

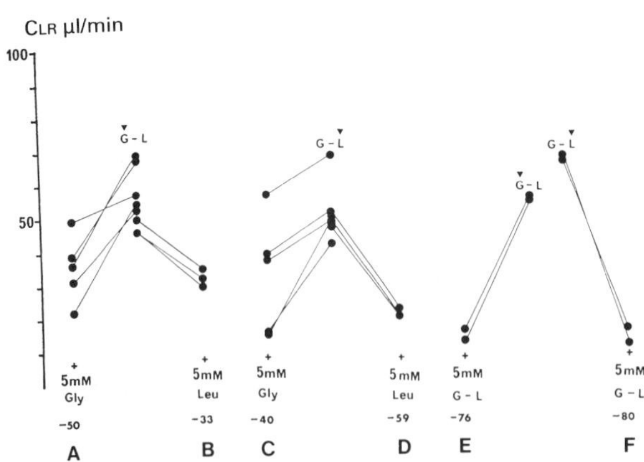

Figure 9.8 Inhibition of [^{14}C-glycyl]-L-leucine (G-L and glycyl-[^{14}C-L-leucine] (G-L), clearance during ventriculocisternal perfusion in the rabbit by either 5 mM glycine, leucine or glycyl-L-leucine. Perfusion was for 150 min; after 75 min perfusion the perfusate was exchanged for one containing exactly the same radioactive concentration of tracer plus a 5 mM concentration of inhibitor. Animals thus act as their own controls. The lines connect data points from single experiments. Average percentage inhibition is shown. Comparison of data points by paired Student's t. (A) $p<0.005$; (B) $p<0.05$; (C) $p<0.025$; (D) $p<0.0005$; (E) and (F) insufficient data points for test

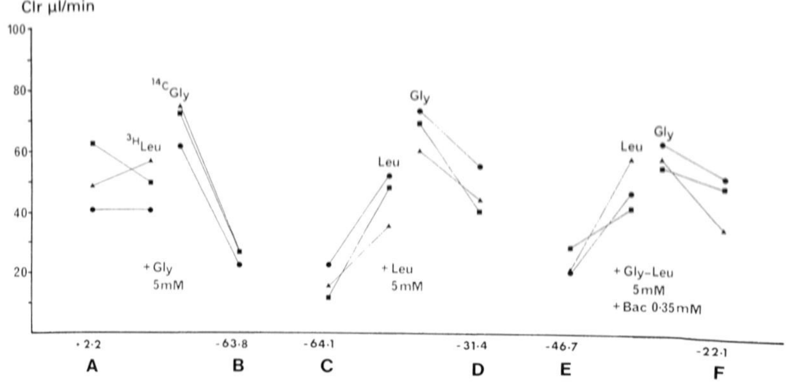

Figure 9.9   Inhibition of [³H]-leucine and [¹⁴C]-glycine clearance by either 5 mM glycine, leucine or glycyl-L-leucine. Details of perfusion as Figure 9.8. Average percentage inhibition is shown. Comparison of data points by paired Student's *t*. (A) ns; (B) *p*<0.005; (D) *p*<0.025; (D) *p*<0.025; (E) *p*<0.05; (F) *p*>0.05<0.1. (The poor level of significance for glycine inhibition by glycyl-L-leucine may be due to a combination of the high clearance value for glycine combined with the observation that glycyl-L-leucine reacts with both the leucine and the glycine transporter. Thus, the available concentration of glycyl-L-leucine available to react with the glycine transporter may be below a fully saturating concentration)

the leucine residue was [¹⁴C]-labelled. The clearance value in microlitres of CSF cleared per minute is the same for each form of the peptide, providing further evidence for a lack of hydrolysis prior to translocation from the ventricles. The clearance from the ventricles of both forms of the peptide are inhibited by both 5 mM glycine and leucine, but a 5 mM concentration of the dipeptide is most effective, producing a 75–80% inhibition. Self-inhibition rarely exceeds this value, as discussed by Begley and Chain (1988), and indicates the probable presence of a non-suppressible, relatively non-specific component to the clearance of small peptides and amino acids which, in the presence of a saturating concentration of competitor, produces a clearance in excess of that expected from the turnover of CSF. Figure 9.9 demonstrates a lack of inhibition of [³H]-leucine clearance by 5 mM glycine, while [¹⁴C]-glycine is effectively self-inhibited at this concentration. On the other hand, leucine at 5 mM concentration is effective at inhibiting the clearance of both tracer leucine and glycine. Figure 9.9 also shows that the glycine and leucine transporters are sensitive to competitive concentrations of glycyl-L-leucine. The interpretation of these results is shown pictorially in Figure 9.10. The simplest model that can be constructed is of two transporters, one primarily transporting glycine and the other leucine. The leucine transporter is BCH sensitive (Begley and Chain, unpublished result) and thus is probably the L-system transporter of Christensen; the glycine transporter appears to be methyl-aminoiosobutyric acid (MeAIB) insensitive (Begley and Chain, unpublished result) and is thus probably system Gly rather than system A.

(MeAIB, the defining substrate for the A-system of Christensen, 1973.) The transporter for glycine will accept leucine but not the other way around, thus accounting for the inhibition of glycine transport by leucine but not vice versa, as shown in Figure 9.9. The dipeptide is acceptable to both the glycine and the leucine transporter; hence the more effective self-inhibition by the dipeptide and less effective inhibition by the two constituent amino acids. This analysis provides strong evidence that, first, the peptide is leaving the ventricles intact and that also it is utilizing the two amino acid transporters primarily directed towards glycine and leucine.

Turning to similar experiments with leucine-enkephalin, the analysis becomes more complex, as more potential competitors are possible. Figure 9.11 shows typical results obtained. Figure 9.11(A,B) shows the clearance of free [^{14}C]-tyrosine, which is competitively inhibited maximally by a 5 mM concentration of tyrosine and to a smaller extent by 5 mM leucine, indicating that the amino acids share a majority of transporters. This observation agrees with the reactivities proposed by Christensen (1973) for these amino acids. Figure 9.11(C,D) shows the inhibition of clearance of [^{3}H-Tyr]-leucine-enkephalin by 5 mM tyrosine and 5 mM BCH. BCH produces a smaller percentage inhibition than does tyrosine, suggesting that leucine-enkephalin may be transported by a number of carriers, only one of which resembles the L-system of Christensen. Tyrosyl-glycyl-glycine, the N-terminal tripeptide of leucine-enkephalin, is almost as effective in suppressing clearance as is tyrosine (Figure 9.11E). Self-inhibition with leucine-enkephalin was performed at concentrations of 0.62–0.72 mM, producing significant inhibition (Figure 9.11F). Finally

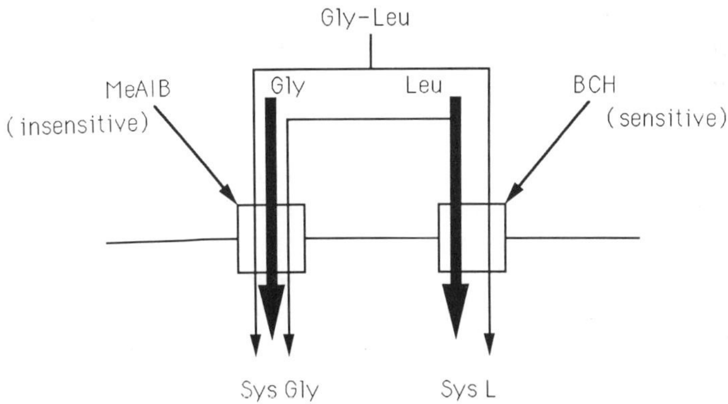

Figure 9.10 Model of glycyl-L-leucine clearance from the cerebral ventricles of the rabbit during ventriculocisternal perfusion. It is suggested that there are two transporters, one of which is BCH sensitive, probably system L, and one of which is MeAIB (methylaminoiso-butyric acid) insensitive, possibly system Gly. There is a low reactivity of leucine with the glycine transporter but not vice versa. Glycyl-leucine on the other hand is able to react with both the glycine and the leucine transporter. Thus, its clearance is competitively inhibited by both glycine and leucine but is most effectively self-inhibited

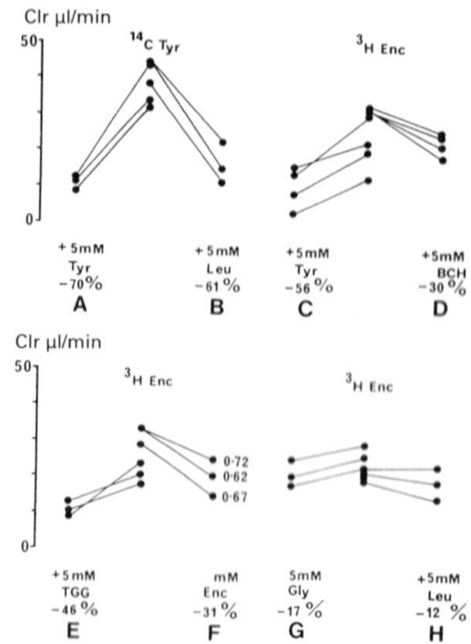

Figure 9.11 Inhibition of [^{3}H-Tyr]-leucine-enkephalin by 5 mM tyrosine, BCH and tyrosyl-glycyl glycine; 0.62, 0.67, 0.72 mM leucine-enkephalin; and 5 mM glycine and leucine. Also inhibition of clearance of [^{14}C]-tyrosine by 5 mM tyrosine and leucine. Details of perfusion as in Figure 9.8. Average percentage inhibition is shown. Comparison of data points by paired Student's t test. (A) $p<0.005$; (B) $p<0.005$; (C) $p<0.025$; (D) $p<0.005$; (E) $p<0.05$; (F) $p<0.01$; (G) $p<0.005$; (H) ns

(Figure 9.11G,H), 5 mM glycine and 5 mM leucine have an insignificant effect on the clearance of leucine-enkephalin, indicating little reactivity with the transporters for these amino acids. These results are interpreted (Begley and Chain, 1988) as shown in Figure 9.12. Leucine-enkephalin is transported by a system primarily directed at the N terminus of the molecule (System N-Tyr), which is tyrosine sensitive and might be a variant of a tyrosine transporter. The inhibitions produced by 5 mM tyrosine and tyrosyl-glycyl-glycine are similar and not significantly different (Begley and Chain, 1988). Leucine-enkephalin produces significant self-inhibition at concentrations of less than 1 mM. There is a small reactivity with the L-system transporter; hence the inhibition produced with BCH. It may well be that further work shows this model to be too simple to fully explain leucine-enkephalin transport. Banks and his collaborators have demonstrated a similar transport mechanism which transports Tyr-MIF 1 and methionine-enkephalin out of the central nervous system (Banks *et al.*, 1986; Banks and Kastin, 1990). This transport system they have termed PTS 1 (protein transport system 1). PTS 1 probably has some affinity for leucine-enkephalin but is not its major transporter; also, PTS 1 does not transport other opiates, including kyotorphin, dynorphin (1–7), β-endorphin and dermorphin. It has some characteristics in common with

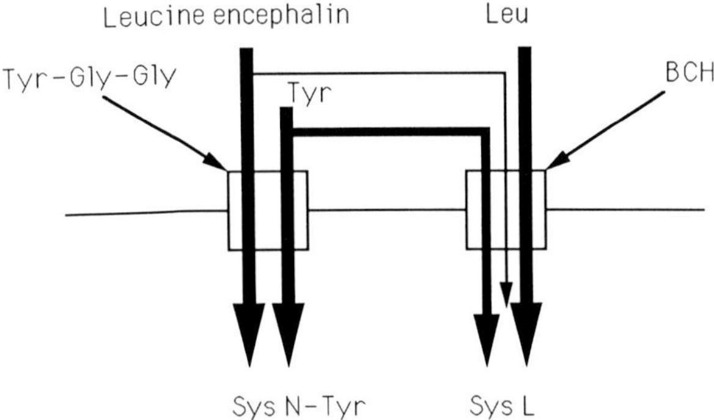

Figure 9.12 Model of [³H-Tyr]-leucine-enkephalin transport from the cerebral ventricles of the rabbit during ventriculocisternal perfusion. Two transporters are proposed, one being the L-system of Christensen, which is BCH sensitive. This system also reacts with tyrosine and, to a lesser extent, leucine-enkephalin. The other transporter is system *N*-Tyr, which transports leucine-enkephalin and tyrosine. It is tyrosyl-glycyl-glycine sensitive and appears to prefer peptides with an N-terminal tyrosine residue. The transporter may be a subtype of the tyrosine transporter. This model will account for a maximal self-inhibition, a strong inhibition by TGG and tyrosine together with a small inhibition by leucine and a minimal inhibition by BCH

system *N*-Tyr described here, in that the presence of an N-terminal tyrosine is a requirement for transport; however, PTS 1 does not appear to be sensitive to inhibiting concentrations of tyrosine (Banks *et al.*, 1986), while system *N*-Tyr is tyrosine sensitive. Thus, there is good evidence for two separate opiate peptide transporting mechanisms out of the central nervous system. As previously mentioned, there are multiple interfaces and multiple carriers for peptides and amino acids in the central nervous system. The work described here utilizes a long-term perfusion technique, while Banks and his group have used an intraventricular injection technique with sampling of brain tissue after a fixed interval. The differences in methodology employed may place different emphases on different transport mechanisms and pathways for these peptides.

TRH, because of its low rate of clearance from CSF, does not allow a model to be constructed by comparison of competitive inhibition data in the same manner as glycyl-L-leucine and leucine-enkephalin. The clearance of the tripeptide TRH from the cerebrospinal fluid was less than half of that of the pentapeptide leucine-enkephalin. It is therefore considered possible that the modifications to the N and C termini of TRH not only confer some protection against hydrolysis in the cerebrospinal fluid, but also reduce the reactivity of the peptide with potential transport mechanisms. One possible transport mechanism for which TRH might have some affinity is that transporting glutamic acid from the cerebrospinal fluid. Hence, the clearance of glutamic acid, pyroglutamic acid (cyclization) and glutamine (amidation) were compared, using the ventriculocisternal perfusion technique (Figure 9.13). The clearance of glutamic acid is $67.9 + 2.5$ μl/min and is effectively suppressed by 73% by the addition of 5 mM glutamic acid to the perfusate. The clearance of glutamine, at $38.9 + 6.1$ μl/min is much less rapid than that of glutamic acid but is still suppressed by some 70% by 5 mM glutamic acid, showing that the tracers are still utilizing substantially the same transporters. Pyroglutamic acid shows a smaller clearance still of $24.8 + 2.3$ μl/min, which is also inhibited by glutamic acid in the perfusate. It is thus concluded that glutamic acid, glutamine and pyroglutamic acid all share common transporters removing them from cerebrospinal fluid. Pyroglutamic acid only has a single carboxy terminal and no free amino terminal, which reduces its reactivity with the transporters relative to glutamic acid, and glutamine has a free amino group and a single carboxy group and its clearance is intermediate between that of pyroglutamic acid and that of glutamic acid, which has a free N amino group and two carboxy terminal groups and the largest clearance value. The clearance of TRH with both N and C termini modified is very small and is equivalent to the non-specific element of clearance observed with other tracers under conditions of maximal carrier inhibition. TRH is thus cleared from cerebrospinal fluid by a combination of bulk flow and the non-specific component.

Thus, not only may TRH and a number of other peptides protect themselves against hydrolysis by modification of the N and C termini, but also these mechanisms may be significant in reducing their reactivity with the major transporting mechanisms that remove peptides from cerebro-

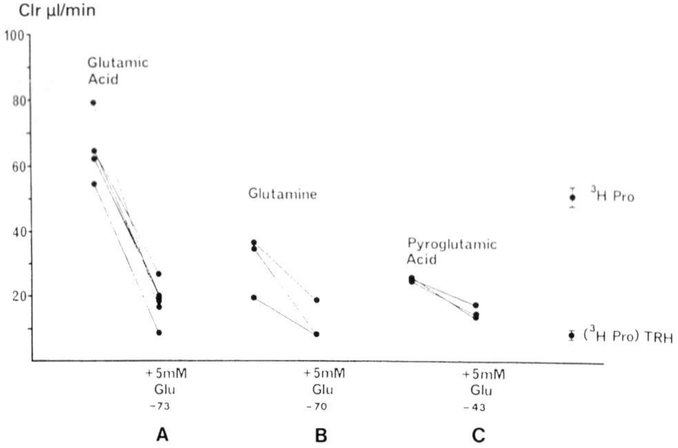

Figure 9.13 Inhibition of glutamic acid, glutamine and pyroglutamic acid clearance from the cerebral ventricles of the rabbit during ventriculocisternal perfusion. The inhibiting amino acid in each case is glutamic acid at 5 mM concentration. Details of perfusion as in Figure 9.8. Average percentage inhibition is shown. Comparison of data points by paired Student's t test. (A) $p<0.001$; (B) $p<0.05$; (C) $p<0.02$. For comparison the mean clearance ($\pm$SEM, $n = 6$) of [³H]-proline and [³H-prolinamide]-TRH is shown

Table 9.4   Clearance of peptides from the cerebrospinal fluid of the rabbit, measured with a ventriculocisternal perfusion technique

	Clearance (μl/min)	T_{50} (min)
Blue dextran (2MD)	0.0	99.0
Insulin	2.6 ± 0.9	80.0
Sucrose	2.6 ± 0.7	79.4
(GSH)$_2$	8.1 ± 1.2	55.9
TRH	8.4 ± 1.2	55.0
Angiotensin II[a]	14.9 ± 3.6	40.9
[^{3}H-Tyr]-DADLE	15.0 ± 2.1	40.8
[^{3}H-Tyr]-leucine-enkephalin[a]	20.0 ± 1.7	34.1
Tyrosine	34.0 ± 2.1	23.4
GSH	35.2 ± 2.5	22.7
[^{14}C-Gly]-Leu	54.2 ± 2.5	16.1
Leucine	58.7 ± 2.7	15.0
Glutamic acid	65.4 ± 1.9	13.7
Glycine	66.7 ± 1.2	13.5

Calculated T_{50} values assume a perfused volume of 1500 μl and a rate of CSF production of 10.5 μl/min.
$T_{50} = \ln2$/rate constant. (Rate constant $= F/V$; $\ln2 = 0.693$.)
(GSH)$_2$ = oxidized glutathione.
GSH = reduced glutathione.
[a] Known peptidase activity in CSF.

spinal fluid, thus increasing their half-lives and activity in the central nervous system.

In conclusion, Table 9.4 illustrates the calculated clearance rates in μl/min over and above that expected from bulk flow, with the calculated half-lives (T_{50}), for a range of peptides studied in our laboratory. Those peptides and tracers with short half-lives are obviously suited to short-term phasic actions within the CNS. With long half-life peptides, their characteristics are more suited to neuromodulatory roles.

Over the long term, peptides with a long half-life and a finite entry into CSF may attain relatively stable CSF concentrations and may represent a physiological set point to which the endocrine system may refer back after peptide hormone levels in the peripheral circulation have been altered by secretion or an increased rate of peripheral utilization or clearance. This set point may be a releasing hormone or the hormone itself.

Also, in the pharmacological investigation and design of drugs and analogues of peptides, with potential CNS actions after peripheral administration, attention must be paid not only to the ease of penetration into the CNS by natural permeability or by utilizing an existing transport mechanism, but also to the subsequent fate and turnover of these molecules once within the brain. The possibilities also arise of pharmacologically manipulating the transport processes regulating peptide concentrations in cerebrospinal fluid to deliberately alter endogenous peptide levels and of designing active peptide analogues which, once within the central nervous system, do not react with these transport mechanisms and are thus not rapidly removed.

ACKNOWLEDGEMENT

The authors would like to thank the Wellcome Trust for financial support and Dr Hugh Davson for his unlimited friendship and generous encouragement in guiding our scientific careers.

REFERENCES

Banks, W. A. and Kastin, A. J. (1990). Peptide transport systems for opiates across the blood–brain barrier. *Am J. Physiol.*, **259** (*Endrocrinol. Metab.* 22), E1–E10

Banks, W. A., Kastin, A. J., Fishman, A. J., Coy, D. H. and Strauss, S. L. (1986). Carrier mediated transport of enkephalins and N-Tyr-MIF-1 across the blood–brain barrier. *Am. J. Physiol.*, **251** (*Endocrinol. Metab.* 14), E477–E482

Begley, D. J. and Chain, D. G. (1981). Clearance of [³H-Tyr]-leucine enkephalin from rabbit cerebrospinal fluid. *J. Physiol (London)*, **319**, 40P–41P

Begley, D. J. and Chain, D. G. (1982). Clearance of glutamic acid, glutamine and pyroglutamic acid from the cerebrospinal fluid of the rabbit; a comparison with thyrotropin-releasing hormone. *J. Physiol. (London)*, **326**, 22P–23P

Begley, D. J. and Chain, D. G. (1988). Transport of enkephalins from the cerebrospinal fluid of the rabbit. In Rakić, Lj., Begley, D. J., Davson, H. and Zloković, B. V. (Eds), *Peptide and Amino Acid Transport Mechanisms in the Central Nervous System*. Macmillan, London, pp. 55–66

Begley, D. J., Davson, H. and Michaelson, I. A. (1980). Clearance of the dipeptide glycyl-L-leucine from rabbit cerebrospinal fluid. *J. Physiol. (London)*, **307**, 83P

Bradbury, M. W. B. and Davson, H. (1964). The transport of urea, creatinine and certain monosaccharides between blood and fluid perfusing the cerebral ventricular system in rabbits. *J. Physiol. (London)*, **170**, 195–211

Christensen, H. (1973). On the development of amino acid transport systems. *Fed. Proc.*, **32**, 19–28

Cserr, H. F. (1971). Physiology of the choroid plexus. *Physiol. Rev.*, **51**, 273–311

Cserr, H. F., Cooper, D. N., Suri, P. K. and Patlak, C. S. (1981). Efflux of radiolabeled polyethylene glycols and albumin from rat brain. *Am. J. Physiol.*, **240** (*Renal Fluid Electrolyte Physiol.* 9), F319–F328

Davson, H., Hollingsworth, J. G., Carey, M. B. and Fenstermacher, J. D. (1982). Ventriculo-cisternal perfusion of twelve amino acids in the rabbit. *J. Neurobiol.*, **13**, 293–318

Davson, H., Kleeman, C. F. and Levin, E. (1963). The blood–brain barrier. In Hogben, A. M. and Lindgren, P. (Eds), *Drugs and Membranes*. Pergamon Press, Oxford, pp. 71–94

Dyer, S. H., Slaughter, C. A., Orth, K., Moomaw, C. R. and Hersh, L. B. (1990). Comparison of the soluble and membrane bound forms of the puromycin-sensitive enkephalin-degrading aminopeptidases from rat. *J. Neurochem.*, **54**, 547–554

Hersh, L. B. (1982). Degradation of enkephalins: the search for an enkephalinase. *Mol. Cell Biochem.*, **47**, 35–43

Hersh, L. B. (1985). Characterization of membrane-bound aminopeptidases from rat brain: identification of the enkephalin degrading aminopeptidase. *J. Neurochem.*, **44**, 1427–1435

Hersh, L. B. and McKelvy, J. F. (1979). Enzymes involved in the degradation of thyrotropin releasing hormone (TRH) and luteinizing hormone (LH-RH) in bovine brain. *Brain Res.*, **168**, 553–564

Iversen, L. L. (1987). Overview: peptides in the nervous system. In Turner, A. J. (Ed.), *Neuropeptides and Their Peptidases*. Ellis Horwood, Chichester, and UCH Verlagsgesellschaft, pp. 3–8

Jackson, I (1980). Significance and function of neuropeptides in cerebrospinal fluid. In Wood, J. (Ed.), *Neurobiology of Cerebrospinal Fluid*, Vol. I. Plenum Press, New York, London, pp. 625–650

Matsui, T., Prasad, C. and Peterofsky, A. (1979). Metabolism of thyrotropin releasing hormone in brain extracts. *J. Biol. Chem.*, **254**, 2439–2445

O'Cuinn, G., O'Connor, B. and Elmore, M. (1990). Degradation of thyrotropin-releasing hormone and luteinizing hormone-releasing hormone by enzymes of the brain. *J. Neurochem.*, **54**, 1–13

Redding, T. W. and Schally, A. V. (1972). On the half life of thyrotropin releasing hormone in rats. *Neuroendocrinology*, **9**, 250–256

Rush, R. S. and Hersh, L. B. (1982). Multiple molecular forms of rat brain enkephalinase. *Life Sci.*, **21**, 445–451

Schelling, P., Ganten, U., Sponer, G., Unger, T. and Ganten, D. (1980). Components of the renin–angiotensin system in the cerebrospinal fluid of rats and dogs with special consideration of the fate of angiotensin II. *Neuroendocrinology*, **31**, 297–308

Schwartz, J. C., Malfroy, B. and De La Baume, S. (1981). Biological inactivation of enkephalins and the role of enkephalin dipeptidyl-carboxypeptidase ('Enkephalinase') as neuropeptidase. *Life Sci.*, **29**, 1715–1740

Segal, M. B. and Zloković, B. V. (1990). *The Blood–Brain Barrier, Amino Acids and Peptides*. Kluwer, Dordrecht, Boston, London

Skidgel, R. A., Defendini, R. and Erdos, E. G. (1987). Angiotensin I converting enzyme and its role in neuropeptide metabolism. In Turner, A. J. (Ed.), *Neuropeptides and Their Peptidases*. Ellis Horwood, Chichester, and UCH Verlagsgesellschaft, pp. 165–182

Turner, A. J. (1987). Endopeptidase-24:11 and neuropeptide metabolism. In Turner, A. J. (Ed.), *Neuropeptides and Their Peptidases*. Ellis Horwood, Chichester, and UCH Verlagsgesellschaft, pp. 183–201

Zloković, B. V., Lipovac, M., Begley, D. J., Davson, H. and Rakić, Lj. (1987). Transport of leucine enkephalin across the blood–brain barrier in the perfused guinea pig brain. *J. Neurochem.*, **49**, 310–315

Zloković, B. V., Lipovac, M., Begley, D. J., Davson, H. and Rakić, Lj. (1988). Slow penetration of thyrotropin-releasing hormone across the blood–brain barrier of an *in situ* perfused guinea pig brain. *J. Neurochem.*, **51**, 252–257

Zloković, B. V., Segal, M. B., Begley, D. J., Davson, H. and Rakić, Lj. (1985). Permeability of the blood–cerebrospinal fluid and blood–brain barriers to thyrotropin-releasing hormone. *Brain Res.*, **358**, 191–199

10
Blood–Brain Barrier Permeability to Peptides and Proteins

Berislav V. Zloković, J. Gordon McComb, Malcolm B. Segal and Hugh Davson

INTRODUCTION

The neurons, glial cells, brain extracellular fluid and cerebrospinal fluid are separated from the blood by the blood–brain and blood–cerebrospinal fluid barriers (Davson, 1976). The blood–brain barrier is well characterized morphologically as a complete and continuous cellular layer of the endothelial cells which are sealed by tight junctions (Brightman, 1977). Normal cell-to-cell communications between astrocytes, pericytes, endothelial cells and surrounding neuropil are essential for the expression of blood–brain barrier phenomena and its homoeostatic mechanisms (Davson and Oldendorf, 1967; Brightman, 1989). Transport, enzymatic and receptor-mediated functions of the blood–brain barrier and blood–cerebrospinal fluid barrier are highly developed, playing a central role in the regulation of the composition of brain extracellular fluid and cerebrospinal fluid. The free movement of circulating hydrophilic substrates from blood to brain extracellular and cerebrospinal fluids is markedly retarded, and it has been accepted that any molecule, above a limiting size, circulating in the blood may gain access to the brain interstitial space only if there is a specific transport system for that molecule localized in the brain capillary endothelium (Oldendorf, 1987; Betz and Goldstein, 1986; Pardridge, 1988). Such transport across the blood–brain barrier must involve three steps: (1) transport across the luminal surface; (2) transfer through the 0.3–0.5-nm-thick endothelial cytoplasm; and (3) transport across the abluminal surface. Specific transport systems at the blood–brain and blood–cerebrospinal fluid barriers have been described for several classes of nutrients (e.g. amino acids, hexoses, monocarboxylic acids), hormones (e.g. thyroid) and water-soluble vitamins (e.g. thiamine) (for review, see Pardridge, 1988).

HISTORY OF THE CONCEPT

Circulating peptides and proteins may come into contact with the brain and cerebrospinal fluid via interfaces located at the luminal side of the blood–brain barrier, the basolateral side of the choroid epithelium and the blood side of the circumventricular organs that are situated outside the barrier (for review, see Segal and Zloković, 1990). The hypothesis that polarity and size of peptide and protein molecules would preclude their rapid transport across the blood–brain barrier has been confirmed by short-term kinetic brain uptake studies (Cornford *et al.*, 1978; Zloković *et al.*, 1985a,b). It has also been suggested that peptide bond may prevent peptides using the L-amino acid transporter at the barrier, and therefore may be responsible for restricting their significant blood-to-brain transfer (Zloković *et al.*, 1983). However, the rapid uptake methods would miss a slow uptake process, as this would be below their level of resolution. Recent studies using the isolated perfused guinea-pig brain preparation over a larger time period have revealed a slow but specific uptake for these molecules at the luminal side of the blood–brain barrier (Zloković *et al.*, 1990b).

The concept of barrier impermeability to proteins has been supported by electron microscopy studies with the plant enzyme horseradish peroxidase (HRP), of which entry into the brain across cerebral capillaries is restricted by the presence of tight junctions and little capacity for vesicular transport (Reese and Karnovsky, 1967; Brightman and Reese, 1969). On the other hand, it has been demonstrated that HRP may penetrate into the brain across the segments of some cerebral arterioles (Westergaard and Brightman, 1973). Other morphological studies on the blood–brain barrier have emphasized that blood-borne native HRP with internalized endothelial surface membrane is directed to endosomes (a prelysosomal compartment) and to secondary lysosomes for eventual degradation without undergoing transendothelial transport (Broadwell *et al.*, 1983). However, considerations that apply to blood–brain transport of exogenous molecules such as HRP and micro-HRP may not necessarily apply to the blood–brain barrier transport of other naturally occurring endogenous proteins and peptides.

Recent work with isolated brain capillaries and autoradiographic analysis of the *in situ* perfused brain has revealed that a number of biologically important proteins and large peptides (e.g. insulin, transferrin, insulin-like growth factors, cationized albumin and immunoglobin G) may bind to capillary endothelial cells; this is followed by subsequent internalization and exocytoses of the engulfed molecules, giving rise to the concept of specialized carrier-, receptor- or absorption-mediated transcytosis (Pardridge, 1986, 1988). The presence of the specific uptake mechanisms for a variety of small neuropeptides (e.g. enkephalins, vasopressin) has been also shown at the luminal side of the BBB by means of an *in situ* brain perfusion technique (Zloković, 1990).

Recent morphological studies with plant lectin wheat germ agglutinin (WGA) conjugated to HRP have offered conclusive morphological evidence for the transcytosis of this blood-borne protein complex through the blood–brain barrier (Broadwell *et al.*, 1988). It has been suggested that transcytosis of blood-borne WGA–HRP across non-fenestrated cerebral endothelia can be attributed to inclusion of the Golgi complex in the transcytotic pathway. However, the physiological/pharmacological significance of the proposed Golgi transcytotic pathway for naturally occurring blood-borne peptides and proteins has yet to be demonstrated.

This review is limited to the behaviour of small neuroactive peptides and immunoglobulin G at the luminal side of the blood–brain barrier and the basolateral face of the choroid epithelium. Our observations are based on two complementary models—i.e. the vascular brain perfusion method in the guinea-pig and the isolated perfused choroid plexus of the sheep.

EXPERIMENTAL MODELS

The experimental models that have been used in our studies have been previously described in detail elsewhere and will be only briefly summarized here.

Vascular Brain Perfusion Model

We have developed a long-term (up to 20 min) vascular brain perfusion model in the guinea-pig to investigate interactions of slowly penetrating neuroactive substances at the blood–brain interface (Zloković *et al.*, 1986). This model allows rigid experimental control of the concentration of peptides and proteins under study, as well as of the composition of arterial inflow. Brain perfusion is carried out *in situ* through the right common carotid artery, which is cannulated by fine polyethylene tubing connected to the extracorporeal perfusion circuit. The contralateral carotid artery is ligated, and both jugular veins are cut to allow drainage of the perfusate. The perfusion medium is an artificial plasma containing sheep red blood cells up to 20%. The surgical procedure in the guinea-pig is reduced compared with the rat model (Takasato *et al.*, 1984), since the cauterization of the right superior thyroid, ophthalmic and pterygopalatine artery is not required. The unique anatomy of the cerebral circulation in *Cavia porcellus* (guinea-pig) has not been found in any mammals other than cavoids (Bugge, 1974). The guinea-pig forebrain blood supply relies on the external carotid artery, since the internal carotid artery normally does not exist, and the flow in the internal ophthalmic artery is reversed under physiological conditions, owing to five anastomotic new branches derived from the proximal part of the external carotid artery and the stapedial system. The circle of Willis is complete, but communication between carotid and vertebral circulations seems to be poor, resulting in the retrograde pressure in the right common carotid artery of between 10 and 15 mmHg. The molecules under study are protected from systemic metabolic influences, owing to functional separation between artificial and vertebral circulations, which has been demonstrated by isotope experiments (Zloković *et al.*, 1989a). HPLC analysis of the radioactive material in the arterial inflow has confirmed that the neuropeptides under study are presented to the blood–brain interface in the intact form.

The physiological (e.g. perfusion pressure, systemic arterial blood pressure, heart performance, spontaneous respiration, cerebral blood flow, acid–base status, brain electrical activity), biochemical (e.g. water and electrolyte regional brain content, ATP and lactate levels, ECP ratio) and morphological (e.g. brain immunogenicity, ultrastructural integrity of the neural tissue in the BBB and non-BBB regions) (Zloković *et al.*, 1986) parameters in the perfused brain remain unchanged in comparison with the normal non-perfused brain.

The unidirectional transfer constant, K_{in}, of the radiolabelled peptide/protein can be determined by multiple time-point/graphic analysis (or single time-point analysis), and Michaelis–Menten parameters for peptide/protein uptake at the luminal side of the BBB can be estimated according to previously published treatments (Pardridge and Mietus, 1982; Patlak *et al.*, 1983; Smith *et al.*, 1987). K_{in} may reflect either cerebrovascular permeability or binding of peptide hormones to the BBB, as in a case of steroid hormone binding and sequestration (Pardridge, 1987). Participation of circumventricular organs and choroid plexuses in overall brain uptake of circulating peptide/protein can be assessed at the same time.

A novel computer-assisted microimaging of the perfused brain has been developed to quantificate the preferential distribution of blood-borne peptide/protein in perivascular brain areas beyond the BBB. This method is based on immunocytochemical analysis of blood-borne peptide/protein in the brain after perfusion. Three-colour processing in red, blue and green spectra permit the quantification of the colour developed during avidin-bitoin–peroxidase reaction.

Chromatographic analysis (HPLC, TLC) of perfused brain tissue can provide additional information regarding peptide metabolism once it has been taken up by the luminal side of the blood–brain barrier.

Isolated Choroid Plexus

The other preparation we have been using in our neuropeptide research is complementary to the previous one, since it permits a study of molecular interactions at the basolateral face of the choroid plexus isolated from the blood–brain barrier, and other circumventricular organs (Deane and Segal, 1985). In this model, all blood vessels of the circle of Willis are ligated, apart from the anterior choroidal arteries, so the auto-perfusion is directed into the plexuses, and the venous drainage is collected from the great vein of Galen. A single-passage paired-tracer indicator dilution method has been applied to the choroid plexus to measure cellular uptake of peptides relative to extracellular space reference markers (Zloković *et al.*, 1985b). The steady-state kinetic analysis used to study sugar transport across the choroid epithelium (Deane and Segal, 1985) has also been applied to the characterization of peptide uptake (Zloković *et al.*, 1988b). It is noteworthy that unidirectional flux of peptide can be measured from several minutes up to 1 h, on the basis of the arteriovenous difference of labelled molecules.

Uptake at the Luminal Side of the Blood–Brain Barrier

By means of the vascular brain perfusion method, a time-dependent progressive linear brain uptake of several neuropeptides has been demonstrated over 20 min (Zloković, 1990b). It has been shown that during relatively short periods of perfusion (1–2 min) it is not possible to distinguish between the extractions of a given peptide from other peptides or metabolically inert polar molecules. However, an extension of the perfusion period up to 20 min has revealed significant differences between slopes of the regression lines for neuropeptides. For example, respective K_{in} values for leucine-enkephalin, arginine-vasopressin, thyrotropin-releasing hormone and delta-sleep inducing peptide were 3.6, 2.8, 1.22 and 0.93 µl/min per gram of the perfused parietal cortex. On the other hand, much lower values were estimated for immunoglobulin G, D-mannitol and dextran (MW 70 000), and the permeability surface area product in the parietal cortex is 0.58, 0.24 and 0.05 $\mu l/ min^{-1}g^{-1}$. All values were statistically significant by ANOVA, indicating the high resolution of this method. It was concluded that cerebrovascular permeability to slowly penetrating neuropeptides is 5–120 times higher than for metabolically inert polar molecules.

The K_{in} values for a tracer in the absence of potential inhibitors and/or competitors represents its maximal permeability surface area product (Smith *et al.*, 1987), and the permeability constant, P, can be calculated on the basis of a known surface area. P values for the above-mentioned neuropeptides, cyclosporin and inert polar molecules have been correlated with their olive oil/water partition coefficients, or reciprocal values of the square root of their molecular weight (Zloković, 1990). This analysis indicated that lipophilicity and molecular weight are not good predictors for cerebrovascular permeability of small peptides.

The first interaction of a circulating neuropeptide at the blood–brain barrier is the luminal receptor site, and a number of experiments have been designed to characterize this event. It has been shown that there is no cross-inhibition between neuropeptides and amino acids at the barrier, and it has been suggested that the N-terminal enzymatic hydrolysis may not be the primary event during peptide interactions at the luminal site. These findings have been supported by a lack of significant inhibition of leucine-enkephalin and arginine-vasopressin uptake in the presence of the amino-peptidase inhibitors bestatin and bacitracin (Zloković *et al.*, 1987, 1989a, 1990a). Our earlier studies indicated that bacitracin reduced leucine-enkephalin brain-uptake index values to the level of intravascular markers, most likely by preventing the action of blood-borne aminopeptidases on the peptide in the bolus (Zloković *et al.*, 1985a). It has been shown that leucine-enkephalin is rapidly degraded *in vitro* (Pardridge and Mietus, 1981), so the lack of this effect in the vascular brain perfusion model may suggest that the enzyme is situated at the abluminal site. It has been demonstrated that the microvascular aminopeptidase M is a membrane-bound protein that contributes to about 1% of the total brain 'enkephalin-ase' activity, but its exact cellular location in the barrier has not been determined (Churchill *et al.*, 1987).

The significant finding in these studies was the dose-dependent self-inhibition that has been demonstrated at the blood–brain barrier for leucine-enkephalin, arginine-vasopressin and delta-sleep inducing factor. The Michaelis constant for leucine-enkephalin was between 35 and 40 μM, which was comparable with that of large neutral amino acids determined during brain perfusion in the rat (Smith *et al.*, 1987). However, the total capacity of this transport system was about three orders of magnitude less than for amino acids. It has been suggested that leucine-enkephalin kinetic parameters are most similar to the blood–brain barrier transport mechan-isms of purine bases and nucleotides. Arginine-vasopressin exhibits a K_m value between 2 and 2.5 μM that is similar to that of the carrier-mediated transport system for thyroid hormones at the barrier (Pardridge, 1988), but with a capacity 35 orders of magnitude lower. It is noteworthy that the diffusion component for both peptides was negligible. In addition, the absence of significant saturable metabolism for peptide fragments has been shown.

The V_1-vasopressinergic receptor antagonist, TMeAVP, but not its V_2 agonist, dDAVP, significantly inhibited vaspressin blood–brain barrier uptake. The affinity of the receptor for TMeAVP was about half of that for the hormone itself (Zloković *et al.*, 1990c). It has been suggested that vasopressin blood–brain barrier receptors either may represent a part of the peptide transport system, as demonstrated for insulin, insulin-like growth factors and transferrin (Pardridge, 1986), or can mediate hormone effects on the barrier, in which case the uptake of blood-borne hormone would be limited primarily to its binding at the blood–brain interface, similar to that observed for steroid hormones (Pardridge, 1987). In contrast to vasopressin, specific inhibitors of the opioid receptors failed to produce any significant effect on leucine-enkephalin transport kinetics, indicating that there is no receptor-mediated component in enkephalin's blood–brain barrier uptake (Zloković *et al.*, 1989a).

The fact that small neuropeptides may be taken up intact by the blood–brain barrier does not rule out the possibility that they may be metabolized in the endothelial cytosol, at the abluminal site or by surrounding neuropil. For example, HPLC analysis of ipsilateral forebrain homogenates indicated that about 70% of vasopressin remained in its intact form between 1 and 3 min, with the percentage of the intact peptide progressively falling with time, so that after 10–15 min, the radioactivity was eluted primarily from the fraction corresponding to radiolabelled phenylalanine. This demonstrates a time-dependent, progressive aminopeptidase degradation of circulating hormone, once it has been transported across the luminal barrier. The possibility that this type of enzymatic degradation may result in formation of a more potent centrally active fragment has been suggested (Burbach *et al.*, 1983).

A saturable mechanism for transport of immunoglobulin G across the blood–brain barrier has also been shown (Zloković *et al.*, 1990), which is in agreement with the recent demonstration of transcytosis of the cationized form of this molecule (Triguero *et al.*, 1989). The Michaelis constant lower than the normal plasma level of immunoglobulin G (Zloković *et al.*, 1989b) suggested that the transport system may be already saturated under physiological conditions.

Uptake at the Basolateral Site of the Choroid Epithelium

Specific peptidergic uptake mechanisms have been demonstrated in the choroid plexus for a number of polypeptides, including melatonin, prolactin and insulin (reviewed by Johanson, 1989). We have also shown that leucine-enkephalin (Zloković *et al.*, 1989b), delta-sleep inducing peptide (Zloković *et al.*, 1989b) and arginine-vasopressin (Zloković *et al.*, 1990c) follow saturable cellular uptake kinetics at the basolateral side of the isolated *in situ* perfused choroid plexus of the sheep. The Michaelis constant for delta-sleep inducing peptide in the choroid plexus of the sheep is 5 nM, indicating high affinity comparable to that found in binding studies with peptide hormones. The arginine-vasopressin system in the choroid plexus of the sheep exhibited a K_m value of 32 nM, and a similar K_m (31 nM) has been estimated in the *in situ* perfused choroid plexus of the guinea-pig (Zloković *et al.*, 1990c). This uptake was strongly inhibited by the V_1 antagonist, TMeAVP, and K_1 values, assuming that the observed inhibitions were purely competitive, ranged from 0.19 to 0.07 μM, for the guinea-pig and sheep choroid plexus, respectively. The V_2 antagonist, dDAVP, and pressinoic acid (1–6 fragment of arginine-vasopressin) produced only weak inhibitions of the peptide uptake in the guinea-pig choroid plexus (but not in the sheep), and K_1/K_m ratios indicated 220 and 310 times lower affinities, respectively, than for arginine-vasopressin. It has been suggested that the membrane mechanism responsible for arginine-vasopressin uptake in the choroid plexus has a binding site with properties similar to those of the V_1 receptor.

CONCLUSION

We should like to conclude that both at the luminal side of the cerebral microvasculature and at the basolateral face of the choroid epithelium there are specific molecular uptake mechanisms that may participate in either transporting peptides across the blood–brain cerebrospinal fluid barriers or mediating their hormonal effects at the blood–brain interface.

ACKNOWLEDGEMENTS

This work was supported by the Wellcome Trust, the Division of Neurosurgery Childrens Hospital of Los Angeles and the British Council.

REFERENCES

Betz, A. L. and Goldstein, G. W. (1986). Specialized properties and solute transport in brain capillaries. *Ann. Rev. Physiol.*, **48**, 241–250

Brightman, M. (1977). Morphology of blood–brain interfaces. *Exp. Eye Res.*, *Suppl.*, 1–25

Brightman, M. W. (1989). The anatomic basis of the blood–brain barrier. In Neuwelt, E. A. (Ed.), *Implications of the Blood–Brain Barrier and Its Manipulation*. Plenum Press, New York, London, pp. 53–78

Brightman, M. W. and Reese, T. S. (1969). Junctions between intimately opposed cell membranes in the vertebrate brain. *J. Cell Biol.*, **40**, 648–677

Broadwell, R. D., Balin, B. J. and Saloman, M. (1988). Transcytotic pathway for blood-borne protein through the blood–brain barrier. *Proc. Natl Acad. Sci. USA*, **85**, 632–636

Broadwell, R. D., Balin, B. J., Saloman, M. and Kaplan, R. S. (1983). Brain–blood barrier? Yes and no. *Proc. Natl Acad. Sci. USA*, **80**, 7352–7356

Bugge, J. (1974). The cephalic arteries of hystriomorph rodents. *Symp. Zool. Soc. Lond.*, **34**, 61–68

Burbach, J. P., Kovasc, G. L. and de Wied, D. (1983). A major metabolite of arginine-vasopressin in the brain is a highly potent neuropeptide. *Science*, **221**, 1310–1312

Churchill, L., Bausback, N. H., Gerritsen, M. E. and Ward, P. E. (1987). Metabolism of opioid peptides by cerebrovascular aminopeptidase M. *Biochim. Biophys. Acta*, **923**, 35–41

Cornford, E. M., Braun, L. D., Crane, P. D. and Oldendorf, W. H. (1978). Blood–brain barrier restriction of peptides and the low uptake of enkephalins. *Endocrinology*, **103**, 1297–1303

Davson, H. (1976). The blood–brain barrier. *J. Physiol. (London)*, **255**, 1–28

Davson, H. and Oldendorf, W. H. (1967). Transport in the central nervous system. *Proc. Roy. Soc. Med.*, **60**, 326–328

Deane, R. and Segal, M. B. (1985). The transport of sugars across the perfused choroid plexus of the sheep. *J. Physiol. (London)*, **328**, 245–260

Johanson, C. E. (1989). Potential for pharmacological manipulation of the blood–cerebrospinal fluid barrier. In Neuwelt, E. A. (Ed.), *Implications of the Blood–Brain Barrier and Its Manipulation*. Plenum Press, New York, London, pp. 223–261

Oldendorf, W. H. (1987). The blood–brain barrier. *Exp. Eye Res.*, *Suppl.*, 177–190

Pardridge, W. M. (1986). Receptor-mediated peptide transport through the blood–brain barrier. *Endocrinol. Rev.*, **7**, 314–339

Pardridge, W. M. (1987). Plasma protein-mediated transport of steroid and thyroid hormones. *Am. J. Physiol.*, **252**, E157–E164

Pardridge, W. M. (1988). Recent advances in blood–brain barrier transport. *Ann. Rev. Pharmacol. Toxicol.*, **28**, 25–39

Pardridge. W. M. and Mietus, L. J. (1981). Enkephalin and blood–brain barrier: studies of binding and degradation in isolated brain microvessels. *Endocrinology*, **109**, 1138–1143

Pardridge, W. M. and Mietus, L. J. (1982). Kinetics of neutral amino acid transport through the blood–brain barrier of the newborn rabbit. *J. Neurochem.*, **38**, 955–962

Patlak, C. S., Fenstermacher, J. D. and Blasberg, R. G. (1983). Graphical evaluation of blood-to-brain transfer constants from multiple-time uptake data. *J. Cereb. Blood Flow Metab.*, **3**, 1–7

Reese, T. S. and Karnovsky, M. J. (1967). Fine structural localization of a blood–brain barrier to exogenous peroxidase. *J. Cell Biol.*, **34**, 207–217

Segal, B. V. and Zloković, B. V. (1990). *The Blood–Brain Barrier, Amino Acids and Peptides*. Kluwer, Dordrecht, Boston, London

Smith, Q. R., Momma, S., Aoyagi, M. and Rapoport, S. I. (1987) Kinetics of neutral amino acid transport across the blood–brain barrier. *J. Neurochem.*, **49**, 1651–1658

Takasato, Y., Rapoport, S. I. and Smith, Q. R. (1984). An *in situ* brain perfusion technique to study cerebrovascular transport in the rat. *Am. J. Physiol.*, **247**, H484–H493

Triguero, D. J., Buciak, J. B., Yang, J. and Pardridge, W. M. (1989). Blood–brain barrier transport of cationized immunoglobulin G: enhanced delivery compared to native protein. *Proc. Natl Acad. Sci. USA*, **86**, 4761–4765

Westergaard, E. and Brightman, M. W. (1973). Transport of proteins across normal cerebral arterioles. *J. Comp. Neurol.*, **152**, 17–44

Zloković, B. V. (1990). *In vivo* approaches for studying peptide interactions at the blood–brain barrier. *J. Control. Rel.*, **13**, 185–202

Zloković, B. V., Begley, D. J. and Chain, D. G. (1983). Blood–brain permeability to dipeptides and their constituent amino acids. *Brain Res.*, **271**, 66–71

Zloković, B. V., Begley, D. J. and Chain-Eliash, D. G. (1985a). Blood–brain barrier permeability to leucine-enkephalin, D-alanine^2-D-leucine^5-enkephalin and their N-terminal amino acid (tyrosine). *Brain Res.*, **336**, 125–132

Zloković, B. V., Begley, D. J., Djuricić, B. and Mitrović, D. M. (1986). Measurement of solute transport across the blood–brain barrier in the perfused guinea-pig brain: method and application to N-methyl-α-aminoisobutyric acid. *J. Neurochem.*, **46**, 1444–1451

Zloković, B. V., Hyman, S., McComb, J. G., Lipovac, M. N., Tang, G. and Davson, H. (1990a). Kinetics of arginine-vasopressin uptake at the blood–brain barrier. *Biochim. Biophys. Acta*, **1025**, 191–198

Zloković, B. V., Lipovac, M. N., Begley, D. J., Davson, H. and Rakić, Lj. (1987). Transport of leucine-enkephalin across the blood–brain barrier in the perfused guinea pig brain. *J. Neurochem.*, **49**, 310–315

Zloković, B. V., Lipovac, M. N., Begley, D. J., Davson, H. and Rakić, Lj. (1988a). Slow penetration of thyrotropin-releasing hormone across the blood–brain barrier of an *in situ* perfused guinea pig brain. *J. Neurochem.*, **51**, 252–257

Zloković, B. V., McComb, J. G., Hyman, S., Perlmutter, L. and Davson, H. (1990b). *Neuroactive Peptides and Amino Acids at the Blood–Brain Barrier: Possible Implications to Drug Abuse.* NIDA Research Monographs, Washington D.C. (in press)

Zloković, B. V., Mackić, J. B., Djuricić, B. and Davson, H. (1989a). Kinetic analysis of leucine-enkephalin cellular uptake at the luminal side of the blood–brain barrier of an *in situ* perfused guinea-pig brain. *J. Neurochem.*, **53**, 1333–1340

Zloković, B. V., Segal, M. B., Begley, D. J., Davson, H. and Rakić, Lj. (1985b). Permeability of the blood–cerebrospinal fluid and blood–brain barriers to thyrotropin releasing hormone. *Brain Res.*, **358**, 191–199

Zloković, B. V., Segal, M. B., Davson, H. and Mitrović, D. M. (1988b). Unidirectional uptake of enkephalins at the blood–tissue interface of the blood–cerebrospinal fluid barrier: a saturable mechanism. *Regul. Peptides*, **20**, 33–44

Zloković, B. V., Segal, M. B., McComb, G. J., Hyman, S., Weiss, M. and Davson, H. (1990c). Kinetics of circulating vasopressin uptake by the choroid plexus. *Am. J. Physiol. (Renal, Fluid and Electrolyte Physiology)* (in press)

Zloković, B. V., Skundrić, D., Segal, M. B., Lipovac, M. N., Mackić, J. B. and Davson, H. (1990d). A saturable mechanism for transport of immunoglobulin G across the blood–brain barrier of the guinea-pig. *Exp. Neurol.*, **107**, 263–270

Zloković, B. V., Susić, V., Davson, H., Begley, D. J., Jankov, R. M., Mitrović, D. M. and Lipovac, M. N. (1989b). Saturable mechanism of delta sleep inducing peptide uptake at the blood–brain barrier of vascular perfused guinea-pig brain. *Peptides*, **10**, 249–254

11
Drainage of Cerebrospinal Fluid During Development and in Congenital Hydrocephalus

Hazel C. Jones

INTRODUCTION

Cerebrospinal fluid is secreted in the cerebral ventricles by the choroid plexuses. It flows through the ventricular system to the foramina in the fourth ventricle and around the subarachnoid space to the drainage sites. Obstructions within the CSF flow pathway, such as developmental abnormalities, haemorrhage and tumours, can cause hydrocephalus by the accumulation of excess CSF in the ventricles. The classical view of CSF drainage is that it takes place through the arachnoid villi in the wall of the cranial venous sinuses (Weed, 1914), although additional routes exist, such as spinal arachnoid villi (Welch and Pollay, 1963) and lymphatic pathways (Weed, 1914; Bradbury and Cole, 1980).

Drainage of the CSF is related to the rate of CSF secretion and to CSF pressure by a general equation where

$$\text{resistance to absorption } (R) = \frac{P \text{ (pressure)}}{Q \text{ (secretion rate or flow)}}$$

and under normal circumstances secretion rate is equal to absorption rate if P is constant. Absorption at the arachnoid villi is dependent on a positive hydrostatic pressure gradient between CSF and venous blood (Davson *et al.*, 1970). The resistance to absorption has been studied in a number of species by perfusion and infusion techniques (Welch, 1975; Blasberg *et al.*, 1981) and in mammals it is inversely related to size, presumably because CSF secretion rate is larger in animals with large ventricles (Davson *et al.*, 1987). During development, increases occur with age in CSF secretion rate (see, e.g., Johanson and Woodbury, 1974) and in CSF pressure (see, e.g., Lorenzo *et al.*, 1981). Thus, corresponding changes in the rate of CSF absorption will be required to maintain equilibrium.

RESISTANCE TO CSF ABSORPTION DURING DEVELOPMENT

The resistance to absorption has been studied in neonatal rats and mice (Jones, 1985; Jones *et al.*, 1987) by use of the constant-rate infusion technique described by Davson *et al.* (1970). With this method, artificial CSF is infused into the lateral ventricle or cisterna magna at infusion rate Q (Figure 11.1) and the CSF pressure increases to a plateau (P_{CSF}), at which time the rate of inflow equals the rate of outflow and for a series of infusion rates resistance to absorption,

$$R = \frac{\Delta P_{CSF} - \Delta P_V}{\Delta Q}$$

where P_V is venous pressure in the dural sinus. If P_V is constant, then

$$R = \frac{\Delta P_{CSF}}{\Delta Q}$$

In newborn rats the resistance to absorption (R) calculated by the second equation is higher by a factor of 8 than in adults and it decreases rapidly in the first 2 weeks after birth (Figure 11.2A). This high R is the same whether measurements are from inside or outside the ventricular system, showing that it is a property of CSF drainage from the subarachnoid space. A similarly high R occurs in newborn mice with a sevenfold decrease by 2 weeks (Figure 11.2B). In this case, however, a proportion of the high resistance is due to a restriction in the outflow from the ventricles, although the decrease with age measured from outside the ventricles in the cisterna

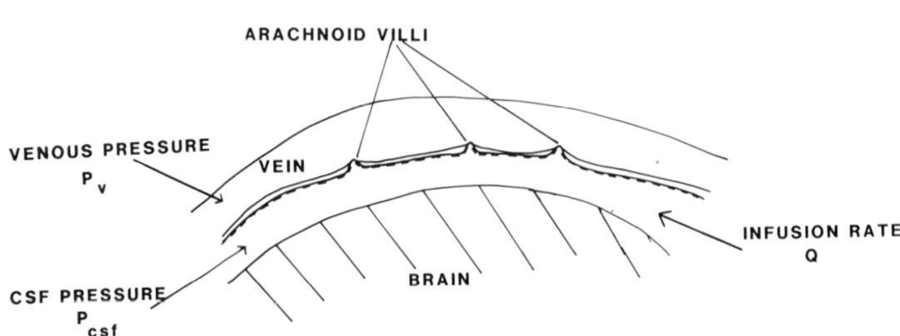

Figure 11.1 The drainage of the cerebrospinal fluid (CSF) through the arachnoid villi depends on the properties of the villi and on the hydrostatic gradient between CSF and dural sinus blood. Infusion of artificial CSF at a series of rates (Q) enables calculation of resistance to drainage as $\Delta P_{CSF}/\Delta Q$. This reflects the properties of the villi if dural sinus pressure (P_V) is constant. If venous pressure increases with CSF pressure, then resistance due to the villi will be $\Delta P_{CSF} - \Delta P_V / \Delta Q$

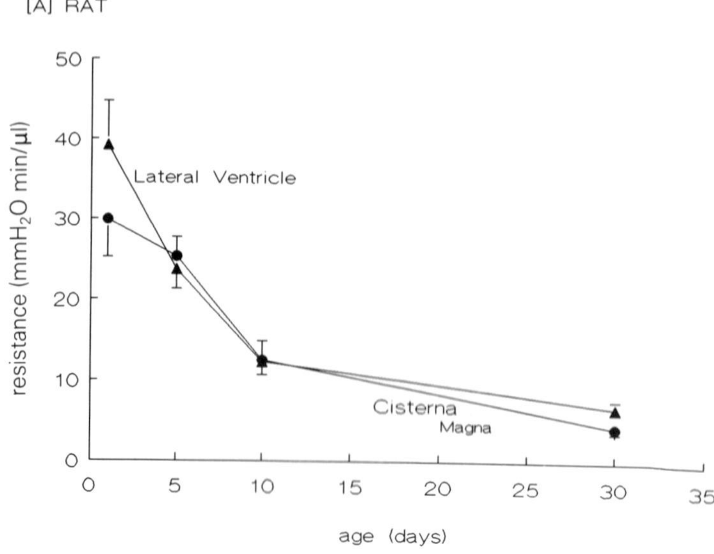

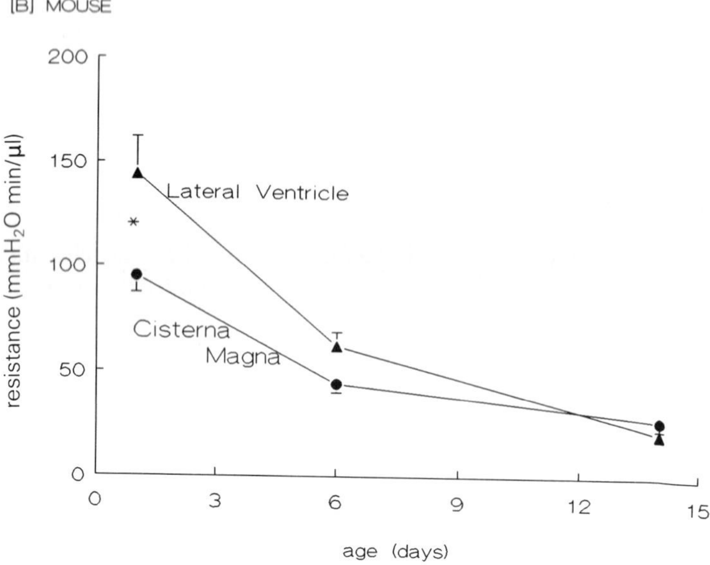

Figure 11.2 Total resistance to absorption ($\Delta P_{CSF}/\Delta Q$) in (A) rats and (B) mice at different postnatal ages. In both species the resistance to drainage is raised in the early postnatal period and falls to adult values within 2–3 weeks after birth. In rats resistance was the same whether measured from the lateral ventricles or the cisterna magna, whereas in mice lateral ventricle resistance was significantly higher than cisterna magna resistance in the youngest age group ($P<0.05$). Points are means $\pm$ SEM, based on 5–10 animals at each age and each site

magna is still fourfold. This suggests that in the young of both these species the venting of excess CSF is more restricted when compared with older animals, and a consequence of this is that the restoration of equilibrium after a disturbance in CSF dynamics will be harder to achieve.

One possible explanation for reduced absorption of CSF from the subarachnoid space in the young animals could be that the outflow sites may be immature and have low conductance properties, or that they may be fewer in number than in older animals, as found in human development (Gomez *et al.*, 1981). Another possible explanation, since CSF drainage depends on the hydrostatic gradient between CSF and venous blood, is that the gradient is less than in older animals. Experimental studies in dogs have shown that sagittal sinus venous pressure does not increase when CSF pressure is raised (Weed and Flexner, 1933; Wright, 1938; Bedford, 1942; Yada *et al.*, 1973) and this is also true for monkeys (Obenchain and Stern, 1973). Cats do show an increase in sinus pressure but it is less than that in the CSF and a positive gradient is maintained (Sahar *et al.*, 1970a). This lack of an effect of intracranial pressure on dural venous pressure is thought to be due to the mechanics of the dura and sinus wall, which are tough and inelastic and do not transmit pressure from the brain compartment. The reverse effect does occur, however: if the dural venous pressure increases for other reasons, such as respiratory distress, there is an increase in CSF pressure, presumably due to an increase of pressure in the small veins which drain into the main sinuses (Guthrie *et al.*, 1970; Yada *et al.*, 1973).

Simultaneous measurements of CSF pressure and dural venous pressure have been made in rats at different ages (Jones and Gratton, 1989a). In rats less than 3 weeks old resting CSF pressure is not significantly different from transverse sinus venous pressure, but subsequently CSF pressure increases with age whereas venous pressure does not (Figure 11.3). Consequently, a clear positive gradient is only seen after 20 days after birth. When CSF plateau pressures are obtained from a series of infusions, there is no effect on venous pressure in adult and 30-day-old rats, similar to previous findings in dogs. In younger rats, however, venous plateau pressures are also obtained, and they increase linearly with CSF pressures but by a smaller amount. The effect is related to age, the largest increase being in the youngest neonate studied (2 days after birth; Figure 11.4). Thus, in the younger rats the pressure gradient from CSF to blood is absent at resting levels and, when the CSF pressure is raised, it is much less than in older rats. Subtraction of the increase in venous pressure ($\Delta P_V/\Delta Q$) from R as calculated by $\Delta P_{CSF}/\Delta Q$ enables R to be partitioned into a villous and venous component (Figure 11.5). The villous component is relatively low throughout the age range and it is possible, therefore, to account for the high resistance in the younger rats by the increase in P_V. This increase in dural sinus venous pressure with CSF pressure in young animals can be

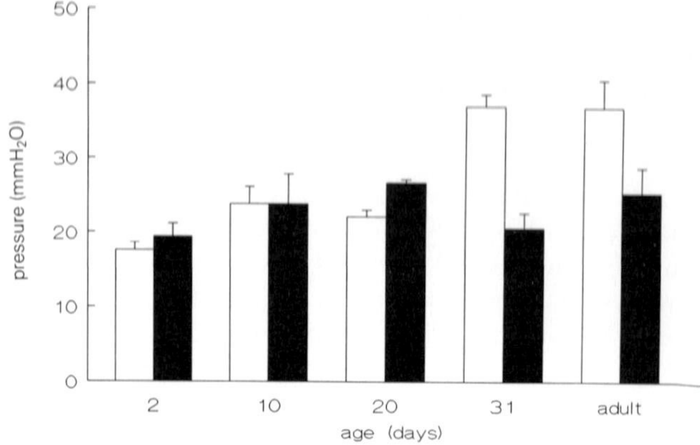

Figure 11.3   Resting CSF pressure and transverse sinus venous pressure in groups of rats at five ages. There is no positive gradient between CSF and venous pressure at the three youngest ages, suggesting that, at normal CSF pressure, drainage through the villi is low or non-existent. Values are means ± SEM; n = 6–7 rats at each age

explained by immaturity of the dura and sinus wall and consequent transmission of pressure from the CSF. It should be pointed out, however, that there could be additional reasons for the high R, such as immaturity of the villi, which would also give a high R and mask the venous pressure effect. Furthermore, it is not known with certainty whether the cranial arachnoid villi actually are the chief route for CSF drainage in young rats. Nevertheless, the rat arachnoid villus has been located in the wall of the superior sagittal sinus and the torcular (Butler *et al.*, 1975, 1983). It is a simpler structure than in larger mammals, and if fixation is carried out at raised CSF pressure, there is an increase in the size and number of pinocytotic vesicles in the endothelial cell cytoplasm, with complete transendothelial channels occurring at very high pressure ($>$650 mmH₂O). An alternative route for fluid transfer could be intercellular channels between the endothelial cells, but Butler *et al.* (1983) found no evidence for this and concluded that transfer of fluid occurs via the cytoplasmic vesicles.

The high resistance to drainage exists for a relatively short time in rats ($<$3 weeks). However, this is a period of rapid brain growth which includes neuronal differentiation, glial cell proliferation, vascularization and myelination (Caley and Maxwell, 1970; Pysh, 1970; Bar, 1980; Korr, 1980). It also covers the main period for growth and modelling of the skull (Massler and Schour, 1951). The question arises as to whether the same high resistance to drainage is present in human infants. Much brain growth is prenatal in man but it continues for an extended period, with dendritic and axonal growth as well as glial cell maturation taking place postnatally (Berry, 1982). A number of groups have measured outflow conductance or

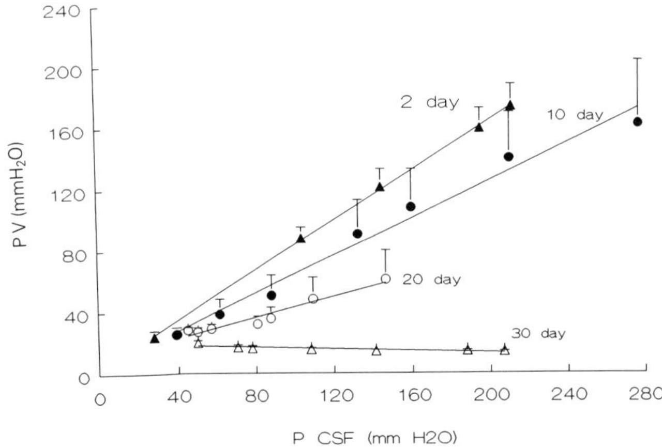

Figure 11.4 The effect of raised CSF pressure on dural sinus venous pressure in rats at four postnatal ages. The increase in venous pressure is less than the increase in the CSF pressure and depends on age: 81%, 61%, 34% and 0% for 2, 10, 20 and 30 day rats, respectively

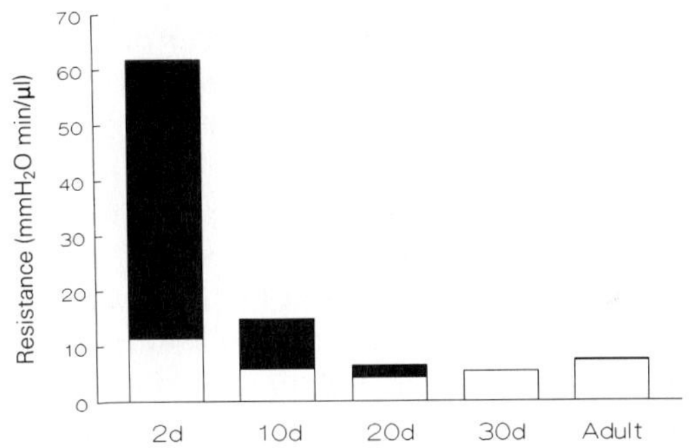

Figure 11.5 The total resistance to absorption ($\Delta P_{CSF}/\Delta Q$) partitioned into a venous component (shaded areas, $\Delta P_V/\Delta Q$) and into a villous component (open areas, $\Delta P_{CSF} - \Delta P_V/\Delta Q$). Most of the high resistance in the younger rats can be accounted for by the increase in venous pressure

resistance to absorption in man by infusion or perfusion methods. However, the data from normal subjects are understandably scarce and most studies span a wide range of ages. Two groups using ventricular perfusion (Cutler *et al.*, 1968; Lorenzo *et al.*, 1970) obtained values for rate of CSF absorption in children up to 14 years old which were similar to those obtained in normal adults (Martins, 1973; Lorenzo *et al.*, 1974). However, this does not exclude the possibility that newborn or premature infants might have a high resistance to drainage as found in young rodents.

CEREBROSPINAL FLUID ABSORPTION IN CONGENITAL HYDROCEPHALUS

The H-Tx rat has inherited congenital hydrocephalus, which first develops in late gestation, is apparent by a domed head at birth and becomes severe within a few weeks of birth. The hydrocephalus is caused by an obstruction of the cerebral aqueduct due to abnormal development of the surrounding brain (Jones and Bucknall, 1988). This rat is an excellent model for hydrocephalus because no artificial induction or mechanical intervention is required to produce the condition and all affected pups have a similar progression of the hydrocephalus. Furthermore, the onset of hydrocephalus occurs before the postnatal period of rapid brain development, allowing investigation of the effect of hydrocephalus on the developing, as opposed to the mature, brain.

In hydrocephalic H-Tx rats resistance to absorption of the CSF ($\Delta P_{CSF}/\Delta Q$) by lateral ventricle infusion is approximately twofold higher than for normal rats at the same age (Figure 11.6A; Jones and Bucknall, 1987). Resistance to absorption by cisterna magna infusion, on the other hand, is close to that for normal rats (Figure 11.6B). This shows that the cause of the hydrocephalus is located in the ventricular system between the two infusion sites and that there is no impairment of the drainage of CSF from the subarachnoid space. It also indicates that drainage of CSF from the lateral ventricles, although slow, is measurable despite the obstructed outflow.

Another question is whether dural sinus pressure behaves similarly in hydrocephalic rats when compared with normal rats at the same age. Hydrocephalus results in enlargement of the head, and skull development is directly affected by the cranial contents (Young, 1959). Furthermore, the fusion of the sutures is delayed and the dural tissue is modified in young rats with induced kaolin hydrocephalus (Young, 1959). Thus, it seems possible that maturation of the tissues separating the CSF and dural sinuses may be delayed. Dural sinus venous pressure has been measured in a group of hydrocephalic H-Tx rats at 10 days after birth (mid-stage hydrocephalus) during infusion into the CSF (Jones and Gratton, 1989b). When compared with normal rats, the hydrocephalic group have an increase in sinus venous pressure which is the same as that in the CSF (Figure 11.7), i.e. 100% of the CSF pressure, whereas in the control group the increase in venous pressure is only 61% of that in the CSF (Figures 11.4, 11.7). This indicates that, without a positive hydrostatic gradient, drainage through the cranial villi is severely impaired in hydrocephalic rats. The most likely explanation for this is that the dura and sinus wall transmit pressure more readily in the hydrocephalics. This may be related to the properties of the sinus wall and the effect on it of the expanding brain. However, it has already been noted that hydrocephalic H-Tx rats have a resistance to

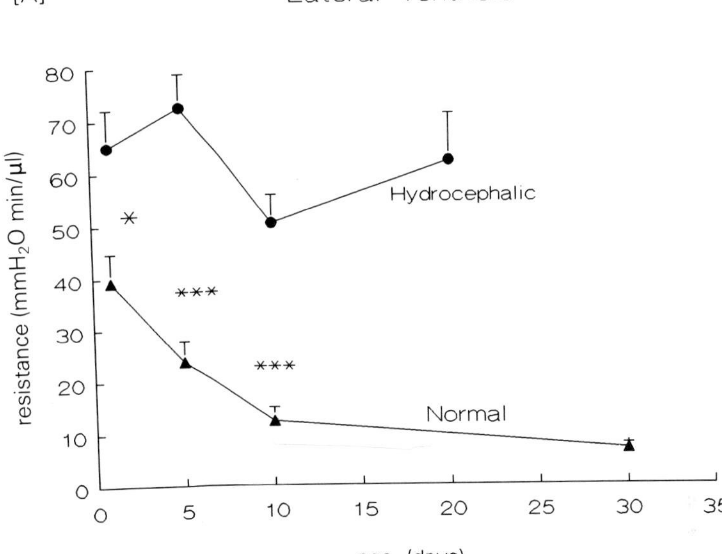

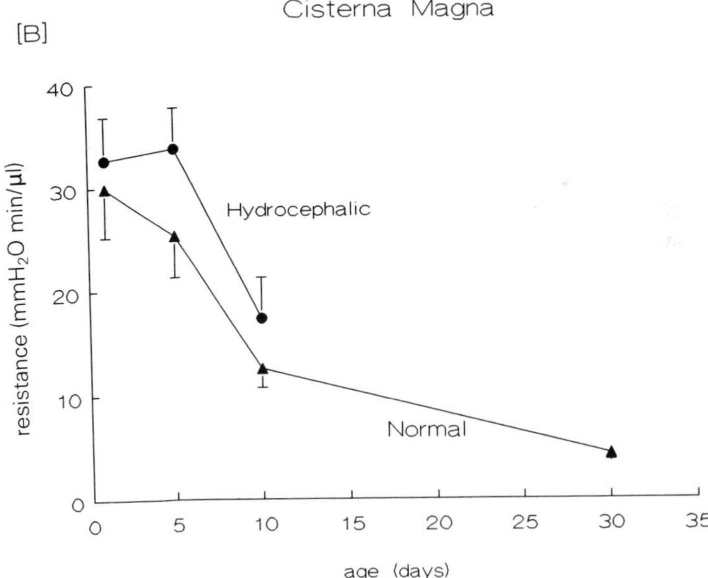

Figure 11.6 Resistance to CSF absorption in hydrocephalic H-Tx rats: (A) from the lateral ventricle, where the hydrocephalic groups have significantly higher resistance than normal rats; (B) from the cisterna magna, where the hydrocephalic groups are not significantly different from normal rats. This indicates that the hydrocephalus is caused by an obstruction in the CSF outflow from the ventricular system

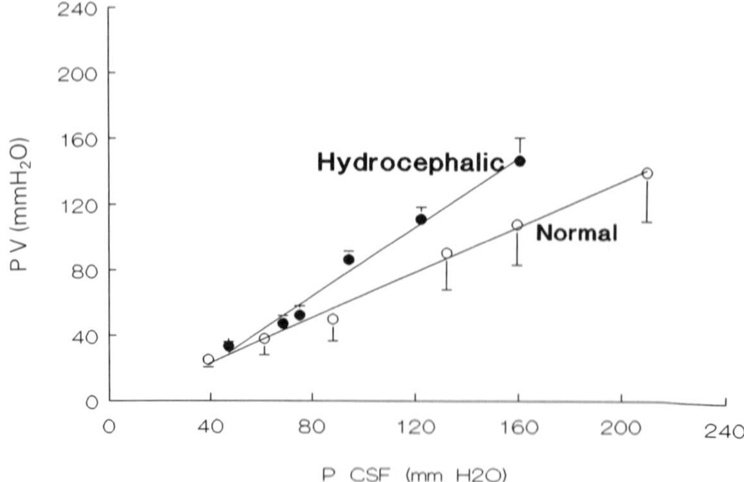

Figure 11.7 The effect of raised CSF pressure on dural venous pressure in normal and hydrocephalic rats at 10 days after birth, corresponding to a mid-stage hydrocephalus. In the hydrocephalic group the venous pressure increases by the same amount as the CSF pressure (100%). In the normal rats the venous pressure response is only 61% of that in the CSF. Thus, the hydrocephalic rats do not have a positive hydrostatic gradient between the two compartments

absorption from the cisterna magna which is similar to that of normal rats. Thus, alternative drainage routes must be operating, such as spinal or lymphatic pathways. This is consistent with other studies on hydrocephalus where, for example, in rabbits with acute hydrocephalus induced by injection of blood, CSF drainage via spinal routes increased and via cranial routes decreased. In another study in cats it was shown that drainage into the sagittal sinus is reduced in hydrocephalus (Sahar *et al.*, 1970b).

Raised dural venous pressure in hydrocephalus may be a common phenomenon. It has been found previously in the sagittal sinus of cats with kaolin hydrocephalus (Sahar *et al.*, 1970a). Similarly, in hydrocephalic dogs both the CSF pressure and the sagittal sinus venous pressure was raised, and in around 50% of dogs sinus pressure was higher than CSF pressure (Shulman *et al.*, 1964). In clinical studies it has been shown that in a group of adults with brain tumours and hydrocephalus 3 out of 12 had increases in sagittal sinus pressure (Martins *et al.*, 1974). In another study on 22 patients with intracranial hypertension, most had a defect in CSF absorption with a small or zero gradient between CSF and sagittal sinus blood and some had total obstruction of the sinuses (Janny *et al.*, 1981). In a group of 15 hydrocephalic infants with various types of hydrocephalus, 11 had a sagittal sinus pressure higher than the CSF pressure and there was a significant positive correlation between the two pressures for the group as a whole (Shulman and Ransohoff, 1965). In another study of 30 infants, over

50% had a sagittal sinus pressure equal to or larger than CSF pressure, and a high proportion showed an increase in sinus pressure when the CSF pressure was raised by infusion (Norrell *et al.*, 1969).

CONCLUSION

In conclusion, experiments have shown that young animals normally have a high resistance to CSF absorption and that this can be accounted for by increases in dural venous pressure. This has implication for CSF drainage in infants because, if resistance to absorption is high, the restoration of equilibrium pressure after a natural disturbance in the system will take longer to accomplish. This might make the CSF system more vulnerable to changes in CSF dynamics. In hydrocephalus infusion into the CSF can help in locating the site of the causative lesion—for example, intra- or extra-ventricular. Furthermore, the accumulated evidence suggests that raised sinus pressure occurs commonly in hydrocephalus. Thus, if CSF pressure rises and there is a consequent increase in venous pressure, CSF absorption could be reduced further and the pathological process accentuated.

REFERENCES

Bar, T. (1980). The vascular system of the cerebral cortex. *Adv. Anat. Embryol. Cell Biol.*, **59**, 1–62

Bedford, T. H. B. (1942). The effect of variations in the subarachnoid pressure on the venous pressure in the superior longitudinal sinus and in the torcular of the dog. *J. Physiol. (London)*, **101**, 362–368

Berry, M. (1982). The development of the human central nervous system. In Dickerson, J. W. T. and McGurk, H. (Eds), *Brain and Behavioural Development*. Surrey University Press

Blasberg, R. G., Johnson, D. and Fenstermacher, J. D. (1981). Absorption resistance of cerebrospinal fluid after subarachnoid haemorrhage in the monkey: effects of heparin. *Neurosurgery*, **9**, 686–691

Bradbury, M. W. B. and Cole, D. F. (1980). The role of the lymphatic system in drainage of cerebrospinal fluid and aqueous humour. *J. Physiol. (London)*, **299**, 353–365

Butler, A. B., Mann, J. D. and Bass, N. H. (1975). Identification of the major site for cerebrospinal fluid efflux in the albino rat. *Anat. Rec.*, **181**, 323

Butler, A. B., Mann, J. D., Maffeo, C. J., Dacey, R. G., Johnson, R. N. and Bass, N. H. (1983). Mechanisms of cerebrospinal fluid absorption in normal and pathologically altered arachnoid villi. In Wood, J. H. (Ed.), *Neurobiology of the Cerebrospinal Fluid*, Vol. II. Plenum Press, New York

Caley, D. W. and Maxwell, D. S. (1970). Development of the blood vessels and extracellular spaces during postnatal maturation of rat cerebral cortex. *J. Comp. Neurol.*, **138**, 31–48

Cutler, R. W. P., Page, L., Galicich, T. and Watters, G. V. (1968). Formation and absorption of cerebrospinal fluid in man. *Brain*, **91**, 707–720

Davson, H., Hollingsworth, J. and Segal, M. B. (1970). The mechanism of drainage of the cerebrospinal fluid. *Brain*, **93**, 665–678

Davson, H., Welch, K. and Segal, M. B. (1987). *The Physiology and Pathophysiology of the Cerebrospinal Fluid*. Churchill Livingstone, Edinburgh

Gomez, D. G., DiBenedetto, A. T., Paves, A. M., Firpo, A., Hersham, D. B. and Potts, D. G. (1981). Development of arachnoid villi and granulations in man. *Acta Anat.*, **111**, 247–258

Guthrie, T. G., Dunbar, H. S. and Karpell, B. (1970). Ventricular size and chronic increased intracranial venous pressure in the dog. *J. Neurosurg.*, **33**, 407–414

Janny, P., Chazal, J., Colnet, G., Irtum, B. and Georget, A.-M. (1981). Benign intracranial hypertension and disorders of CSF absorption. *Surg. Neurol.*, **15**, 168–174

Johanson, C. E. and Woodbury, D. M. (1974). Changes in CSF flow and extracellular space in the developing rat. In Vernadakis, A. and Weiner, N. (Eds), *Drugs and the Developing Brain*. Plenum Press, New York

Jones, H. C. (1985). The cerebrospinal fluid pressure and resistance to absorption during development in normal and hydrocephalic mutant mice. *Exp. Neurol.*, **90**, 162–172

Jones, H. C. and Bucknall, R. M. (1987). Changes in cerebrospinal fluid pressure and outflow from the lateral ventricles during development of congenital hydrocephalus in the H-Tx rat. *Exp. Neurol.*, **98**, 573–583

Jones, H. C. and Bucknall, R. M. (1988). Inherited prenatal hydrocephalus in the H-Tx rat: a morphological study. *Neuropathol. Appl. Neurobiol.*, **14**, 263–274

Jones, H. C., Deane, R. and Bucknall, R. M. (1987). Developmental changes in cerebrospinal fluid pressure and resistance to absorption in rats. *Dev. Brain Res.*, **33**, 23–30

Jones, H. C. and Gratton, J. A. (1989a). The effect of cerebrospinal fluid pressure on dural venous pressure in young rats. *J. Neurosurg.*, **71**, 119–123

Jones, H. C. and Gratton, J. A. (1989b). The drainage of cerebrospinal fluid in hydrocephalic rats. *Z. Kinderchir.*, **44**, Suppl. I, 14–15

Korr, H. (1980). Proliferation of different cell types in the brain. *Adv. Anat. Embryol. Cell Biol.*, **61**, 1–70

Lorenzo, A. V., Bresnan, M. J. and Barlow, C. F. (1974). Cerebrospinal fluid absorption deficit in normal pressure hydrocephalus. *Arch. Neurol.*, **30**, 387–393

Lorenzo, A. V., Page, L. K. and Watters, G. V. (1970). Relationship between cerebrospinal fluid formation, absorption and pressure in human hydrocephalus. *Brain*, **93**, 679–692

Lorenzo, A. V., Welch, K., Conner, S., Black, P. and Dorval, B. (1981). Changes in the water and electrolyte pattern of the brain and of the intracranial pressure during development in rabbits. *Z. Kinderchir.*, **34**, 410–415

Martins, A. N. (1973). Resistance to drainage of cerebrospinal fluid. Clinical measurement and significance. *J. Neurol. Neurosurg. Psychiat.*, **36**, 313–318

Martins, A. N., Kobrine, A. I. and Larsen, D. F. (1974). Pressure in the sagittal sinus during intracranial hypertension in man. *J. Neurosurg.*, **40**, 603–608

Massler, M. and Schour, I. (1951). The growth pattern of the cranial vault in the albino rat as measured by vital staining with alizarin red 'S'. *Anat. Rec.*, **110**, 83–101

Norrell, H., Wilson, C., Howieson, J., Megison, L. and Bertan, V. (1969). Venous factors in infantile hydrocephalus. *J. Neurosurg.*, **31**, 561–569

Obenchain, T. G. and Stern, W. E. (1973). Continuous pressure monitoring in experimental hydrocephalus. I. The dynamics of acute ventricular obstruction. *Arch. Neurol.*, **29**, 287–294

Pysh, J. J. (1970). Mitochondrial changes in rat inferior colliculus during postnatal development: an electron microscopic study. *Brain Res.*, **18**, 325–342

Sahar, A., Hochwald, G. M. and Ransohoff, J. (1970a). Cerebrospinal fluid and cranial sinus pressures. Relationship in normal and hydrocephalic cats. *Arch. Neurol.*, **23**, 413–418

Sahar, A., Hochwald, G. M. and Ransohoff, J. (1970b). Passage of cerebrospinal fluid into cranial venous sinuses in normal and experimental hydrocephalic cats. *Exp. Neurol.*, **28**, 113–122

Shulman, H. and Ransohoff, J. (1965). Sagittal sinus venous pressure in hydrocephalus. *J. Neurosurg.*, **23**, 169–173

Shulman, K., Yarnell, P. and Ransohoff, J. (1964). Dural sinus pressure in normal and hydrocephalic dogs. *Arch. Neurol.*, **10**, 575–580

Weed, L. H. (1914). Studies on cerebrospinal fluid. III. The pathways of escape from the subarachnoid spaces with particular reference to the arachnoid villi. *J. Med. Res.*, **31**, 51–91

Weed, L. H. and Flexner, L. B. (1933). The relations of the intracranial pressures. *Am. J. Physiol.*, **105**, 266–272

Welch, K. (1975). The principles of physiology of the cerebrospinal fluid in relation to hydrocephalus including normal pressure hydrocephalus. *Adv. Neurol.*, **13**, 247–332

Welch, K. and Pollay, M. (1963). The spinal arachnoid villi of the monkeys *Cercopithecus aethiops sabaeus* and *Macaca irus*. *Anat. Rec.*, **145**, 43–48

Wright, R. D. (1938). Experimental observations on increased intracranial pressure. *Aust. New Zeal. J. Surg.*, **7**, 215–235

Yada, K., Nakagawa, Y. and Tsuru, M. (1973). Circulatory disturbances of the venous system during experimental intracranial hypertension. *J. Neurosurg.*, **39**, 723–729

Young, R. W. (1959). The influence of cranial contents on postnatal growth of the skull in the rat. *Am. J. Anat.*, **105**, 383–415

12
Development of the Blood–Brain Barrier to Macromolecules

N. R. Saunders

INTRODUCTION

It is perhaps not necessary to remind readers of a book on the blood–brain barrier that the concept of the blood–brain barrier is now nearly 100 years old and the earliest experiments on the development of the blood–brain barrier were published over 60 years ago (Wislocki, 1920; Behnsen, 1927). The original idea of a barrier protecting the brain from substances in the circulating blood stemmed, of course, from the dye injection experiments of Ehrlich and others (see, e.g., Ehrlich, 1885). Today the blood–brain barrier has a much more general meaning and the term is used to describe a series of mechanisms that control the internal environment of the brain. These mechanisms not only include the exclusion or near-exclusion of dyes (bound to protein) and other macromolecules, but also include more active mechanisms that control the transfer of various materials (ions, amino acids, glucose, vitamins) into or out of the brain. The nature of the barrier to macromolecules is principally the presence of tight junctions between cerebral endothelial and between choroid plexus epithelial cells (Brightman and Reese, 1969). This barrier is fundamental to all the other barrier mechanisms, since, in the absence of a diffusion restraint, the characteristic gradients, for example, for electrolytes between CSF and brain ECF on the one hand and plasma on the other hand, could not be sustained.

This chapter will review briefly the historical evidence concerning the permeability of the blood–brain barrier to macromolecules in the developing brain. The main part of the chapter will be divided into three further sections: (1) the morphology of barriers to macromolecules in the developing brain; (2) the permeability of brain barriers to macromolecules of the size of plasma proteins (>2.5 nm); and (3) the permeability of brain

128

barriers to macromolecules of the size of inulin (1.3 nm) and less.

Current evidence shows that tight junctions at both barrier sites (blood–brain and blood–CSF) develop very early indeed and that the brain itself is 'protected' from the high concentration of protein in the extracellular/plasma environment of the rest of the embryo or fetus from the very earliest stages of development. In addition, there is a barrier present in the developing brain which is not found in the adult. This is at the interface between CSF and brain extracellular fluid (ECF) and takes the form of a membrane specialization between the cells of the neuroependyma that line the cerebral ventricles ('strap junctions': see Møllgård *et al.*, 1987). These junctions prevent the intercellular passage of proteins from the high concentrations present in fetal CSF (Dziegielewska and Saunders, 1988) to the brain ECF.

Like the tight junctions between cerebral endothelial cells, the tight junctions between choroid plexus epithelial cells are also found very early in brain development (Møllgård *et al.*, 1976; Møllgård and Saunders, 1986). However, the well-formed intercellular blood–CSF barrier to macromolecules is by-passed by a *trans*cellular mechanism that selectively transfers proteins from blood to CSF and is largely if not entirely responsible for the very high concentration of protein in fetal CSF (see below).

Thus, the originally described barrier to dyes, which is in fact a barrier to dye bound to protein (Tschirgi, 1950) is well formed in the fetus and the commonly stated belief that 'the blood–brain barrier in the fetus and newborn is immature or highly permeable' (see, e.g., Ganong, 1989) is incorrect when considering macromolecules of the size of plasma proteins or molecules that bind to such proteins.

It does seem, however, that both the blood–brain and blood–CSF barriers in the developing brain are more permeable to macromolecules much smaller than proteins (e.g. inulin). This was probably first observed as long ago as the 1920s by Stern and her colleagues (Stern and Rapoport, 1928; Stern *et al.*, 1929; see below) but this seems to have gone largely unnoticed until recently.

EARLY EXPERIMENTS ON BARRIER PERMEABILITY IN IMMATURE ANIMALS

Behnsen (1927) is generally credited with having been the first to try dye injection experiments in immature animals, although Wislocki (1920) published results of such experiments in fetal cats several years earlier and come to the opposite conclusion. Behnsen's experiments were on postnatal rats and were interpreted (and widely quoted) as demonstrating immaturity of the blood–brain barrier. Wislocki (1920), Behnsen (1927), Penta

(1932) and Stern and colleagues (e.g. Stern *et al.*, 1929) published a series of dye injection experiments in the 1920s and 1930s with apparently different results. Thus, Behnsen and Penta reported greater staining of the immature brain with dyes such as trypan blue than occurs in adults. In contrast, Wislocki (1920), Stern and Rapoport (1928) and Stern *et al.* (1929) found no such staining at any age in any of several species investigated, including cat and guinea-pig fetuses and neonatal rats, mice, cats, dogs and rabbits. In fact, neither Behnsen nor Penta reported 'blue brains' in the sense of a generalized leak of dye from blood into brain. Behnsen stated, for example, that the region of the brain stained was an extension of the rather larger area that normally stains in the adult owing to the lack of a barrier—i.e. hypothalamus, median eminence, etc. This appearance would have been partly due to the fact that the hind and mid parts of the brain develop in advance of the cortex which so dominates the appearance of the adult brain; thus, proportionately speaking, the area of staining one expects to see in the adult brain would appear greater in the immature brain. However, a more important factor relates to the experimental conditions used and, in particular, the amount of dye injected. This was enormous. For example, Penta (1932) injected dye over several days to such an extent that many of his animals died from the toxic effects of the dye. Stern appears to have been well aware of this problem: 'Ces recherches avaient en outre établi que, si la résistance de la barrière chez le nouveau-né etait considerablement plus failble vis-à-vis des substances cristalloides telles que le ferrocyanure ou l'iodure, sa résistance à l'égard des colloides tels que le bleu trypan et le rouge Congo ne différait pas de celle de l'adulte. . . . Ainsi, chez le lapin nouveau-né, le passage du bleu trypan ne s'observe que lorsque la quantité injectée dépasse 3 à 4 gr. par kilogramme d'animal; le passage du rouge Congo ne se mainifeste que lorsque la quantité injectée dépasse 3 gr. par kilogramme d'animal. . . . Il faut atteindre des doses massives dépassant de 30 à 40 fois les doses utilisées normalement dans nos experiences, pour provoquer l'apparition de ces colorants dans le liquide céphalorachidien. Il faut tenir compte des doses lorsqu'il s'agit d'apprécier les résultats obtenus par les divers auteurs.' (Stern *et al.*, 1929). Unfortunately Stern's concerns about the need to use moderate amounts of dye appear to have been largely ignored. The modern version of an Ehrlich dye experiment is the injection of horseradish peroxidase solution. This protein (molecular radius 2.5–3.0 nm) is excluded from the adult brain and CSF (Brightman and Reese, 1969). In recent years several papers have been published describing the penetration of HRP into immature brain and/or CSF. However, most of these papers do not comply with Stern's reasonable strictures on the need to use moderate amounts of marker. Table 12.1 shows estimates of the volume injected compared with the size of animal and estimated circulating blood volume. The likely effects on circulating plasma protein concentra-

Table 12.1 Recent experiments in which horseradish peroxidase (HRP) was injected into immature animals as a test of blood–brain barrier integrity. All claimed that the results suggested that the barrier (to protein) is immature. The authors stated that injection volume has been compared with estimates of circulating blood volume (allowing a factor of 2–3 for placental blood volume in the case of fetuses, although it is doubtful whether, under the conditions of the experiments, much mixing with fetal placental blood would have occurred). The effect of the amount of HRP on the plasma protein concentration is also estimated

Authors	Species	Age (days)	Body weight (g)	Injection vol.	Effect on plasma protein concentration
Wakai and Hirokawa (1981)	chick	E9	1.56	5–10% blood vol.	2 × increase
Risau et al. (1986)	mouse	E13	0.083	1–3 × blood vol.	1% increase
Vorbrodt et al. (1986)	mouse	newborn	1.4	=total blood vol.	2–3 × increase
Stewart and Hayakawa (1987)	mouse	E15	0.26	not stated	4 × increase

tion are also indicated. These changes scarcely seem to be within what might be considered to be reasonable physiological limits.

Stern herself reported (see, e.g., Stern *et al.*, 1929) that provided that reasonable amounts of dye were used, no dye penetrated into CSF or brain at any age in various species investigated. This result was confirmed in the 1950s by several workers (e.g. Grazer and Clemente, 1957). It has also been shown for albumin (Olsson *et al.*, 1968) and HRP (Tauc *et al.*, 1984) in studies in which more moderate amounts were injected into rodent fetuses that the injected protein does not cross the blood–brain barrier in the immature brain. Similar results were obtained with injection of HRP and human protein in newborn tammar wallabies, which are at an extremely immature stage of brain development (Dziegielewska *et al.*, 1988).

Stern appears to have made another important and original observation that has gone largely unnoticed—namely, that there is a difference in the permeability properties of the developing brain with respect to macromolecules of smaller molecular weight (Stern used the term 'crystalloid', e.g. sodium ferrocyanide) when compared with what she called 'colloids'— dyes that were subsequently shown to bind to plasma proteins. Her use of the term 'colloid' is interesting because of its use in relation to plasma proteins (e.g. colloid osmotic pressure) but it is not clear from her papers whether she realized that dyes such as trypan blue bind to plasma proteins, as was clearly shown much later by Tschirgi (1950). Stern showed that, whereas the immature brain was impermeable to 'colloid' dyes, it was much more permeable to 'crystalloids' than in the adult brain (Stern and Peyrot, 1927; Stern *et al.*, 1929). This qualitative observation is a very early forerunner of much later quantitative experiments using radioactively labelled molecules of similar molecular size—sucrose and inulin (Ferguson and Woodbury, 1969; Dziegielewska *et al.*, 1979; see below).

THE MORPHOLOGY OF BARRIERS TO MACROMOLECULES IN THE DEVELOPING BRAIN

The Blood–Brain, Cerebral Endothelial, Barrier

Nature performs a barrier experiment that makes it unnecessary to inject fetuses with either noxious dyes or a plant enzyme (HRP). Stern (1934) appears to have been aware of this, at least in relation to the state of barrier function in adult animals. She comments that, in studying the functional state of the blood–brain barrier, it is important not only to examine substances introduced into the blood, but also to consider the large number of substances occurring naturally in the circulating blood which are also found in the CSF. Thus, if we consider the fetus with this in mind, beginning in the yolk sac and subsequently in the liver, the embryo and later the fetus synthesize a number of proteins that circulate in the plasma: for example, α-fetoprotein, transferrin and albumin. If one cuts sections of immature brain and stains them for one of these proteins, using a suitable immunocytochemical method, then it is possible to determine whether cerebral blood vessels are 'leaky' to protein at any stage of their development. Such a section is illustrated in Figure 12.1. There is a clear immunoprecipitate of protein in CSF within the cerebral ventricles and in plasma in blood vessels. The mesenchymal tissue around the brain is also stained. The endothelial cells themselves appear to contain the plasma protein, but there is no evidence of protein in the extracellular space around the vessels or elsewhere in the brain. In spite of the stained protein in CSF, there is also no evidence of protein passing between the cells of the neuroependyma lining the cerebral ventricles (see section on CSF–brain, neuroepithelial, barrier, below). This lack of staining of the extracellular space of the immature brain for proteins that occur naturally in the circulating blood is in contrast to the results of many experiments involving injection of dyes or HRP, as discussed above (see also Table 12.1).

This lack of penetration of endogenous protein into the immature brain is due to presence of tight junctions between cerebral endothelial cells.

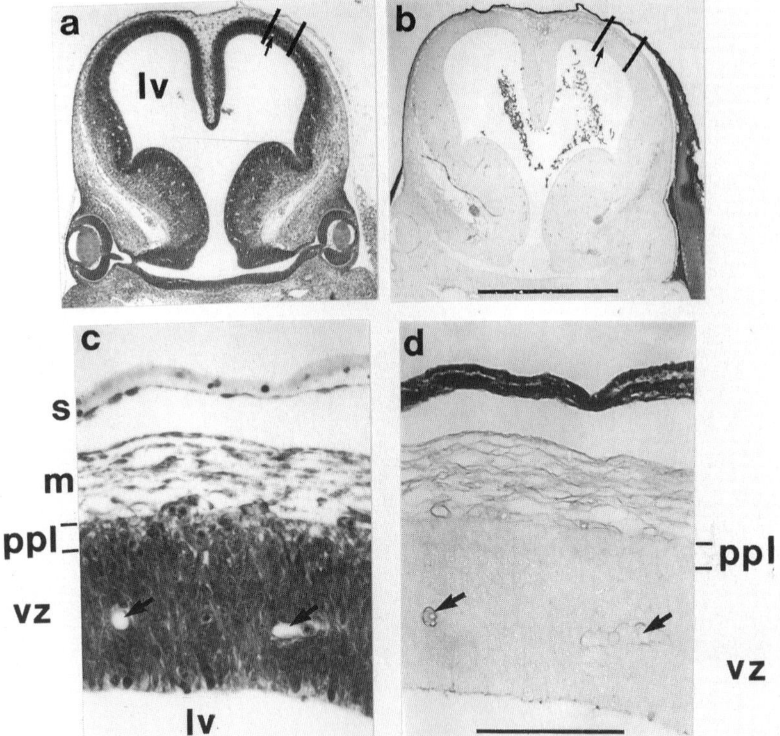

Figure 12.1   Micrographs of fetal rat forebrain at E14 cut in coronal section. (a) Low-power and (c) high-power, toluidine blue stained to show general structure. Note that the wall of forebrain vesicle is at a very early stage of development. It consists of only a wide inner ventricular zone (VZ) containing dividing cells and a narrow outer primordial plexiform layer (ppl) or marginal zone. There are occasional blood vessels within the substance of the developing brain (arrows). (b) Low-power and (d) high-power, adjacent section to (a), stained for plasma proteins, using an antiserum to rat plasma. Arrows indicate blood vessels. Note staining of cerebral endothelial cells but lack of staining for plasma proteins within brain parenchyma. Note precipitated CSF in lateral ventricles (LV), also extracerebral staining of embryonic skin (s) and mesenchymal tissue (m) outside brain. Parallel bars in (a) and (b) indicate regions illustrated at higher power in (c) and (d). Upper bar = 1.0 mm in (a) and (b). Lower bar = 0.1 mm in (c) and (d).

This has been demonstrated in a number of different species (Møllgård and Saunders, 1975, 1986; Saunders and Møllgård, 1984). However, the evidence from the low-power immunocytochemical staining is perhaps more convincing, because with this technique a wide area of brain can be rapidly examined, whereas with the freeze–fracture technique (Figure 12.2) it is difficult to obtain adequate fractures through the endothelial cell-tight junctions and therefore only a relatively few vessels can be examined. Similarly, with thin-section transmission EM only a few vessels can be examined, and unless they are studied in serial section, one cannot know whether the junction is complete.

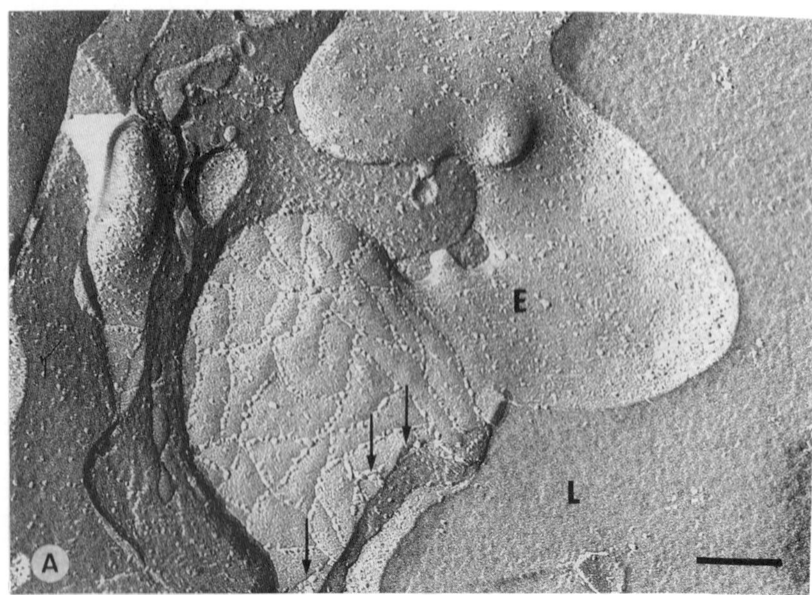

Figure 12.2 Freeze–fracture micrographs of tight junctions between cerebral endothelial cells: (A) capillary in adult rat neocortex; (B) capillary in fetal rabbit neocortex. Arrows indicate tight junctional strands. E = E face; L = lumen; bars = 0.2 μm. Original micrographs kindly supplied by Professor K. Møllgård

The Blood–CSF, Choroid Plexus Epithelial, Barrier

Here the problem is rather different from that of the blood–brain barrier. As indicated in the previous section and Figure 12.1, staining for a naturally occurring plasma protein such as α-fetoprotein in sections of the developing brain reveals that it is present in the CSF, probably in a rather high concentration (as is confirmed by direct measurement; see Dziegielewska and Saunders, 1988, and section below). Does this mean that, in contrast to the well-formed blood–brain barrier to protein, the blood–CSF barrier to large macromolecules is not yet developed in the immature brain? Extensive thin-section and freeze–fracture EM studies of the choroid plexus in immature brains show that tight junctions are present between the epithelial cells of the choroid plexus from as early as the choroid plexus is apparent (Figure 12.3). Substantially more evidence on the completeness of tight junctions in the developing choroid plexus is available, because the junctions are easier to fracture (Møllgård and Saunders, 1975, 1986; Møllgård *et al.*, 1976; Tauc *et al.*, 1984). In addition, sufficient information was available to make a quantitative study of the dimensions and complexity of tight junctions during fetal development, as is shown in Table 12.2, taken from Møllgård *et al.* (1976).

If, as appears to be the case, choroid plexus tight junctions are well-developed, even very early in brain development, the presence of large concentrations of proteins (most if not all of which are immunologically identical with those in plasma) needs to be explained. Choroid plexus cells contain a number of plasma proteins (Jacobsen *et al.*, 1982a,b) and there is experimental evidence for protein transfer, probably across the choroid plexus epithelial cells by an intracellular route, in the immature brain (see section on the blood–CSF, choroid plexus epithelial, barrier, p. 140). The choroid plexus in the developing brain has also been shown to synthesize a number of plasma proteins (Dziegielewska *et al.*, 1984; Saunders, 1984; Thomas *et al.*, 1988, 1989); this does not seem to be a major contributor to the high concentration of protein in fetal CSF, since at the ages studied most if not all of the protein in CSF appears to originate from the plasma (Dziegielewska *et al.*, 1980a,b, 1989b; Habgood, 1990); however, no one has yet studied transfer at the peak of CSF protein concentration in the immature brain.

The CSF–Brain, Neuroepithelial, Barrier

This is perhaps the most unexpected finding from among recent studies of barriers in the developing brain (Fossan *et al.*, 1985; Møllgård *et al.*, 1987). As mentioned above, the high concentration of protein in CSF does not seem to be reflected in extracellular staining for such proteins within the

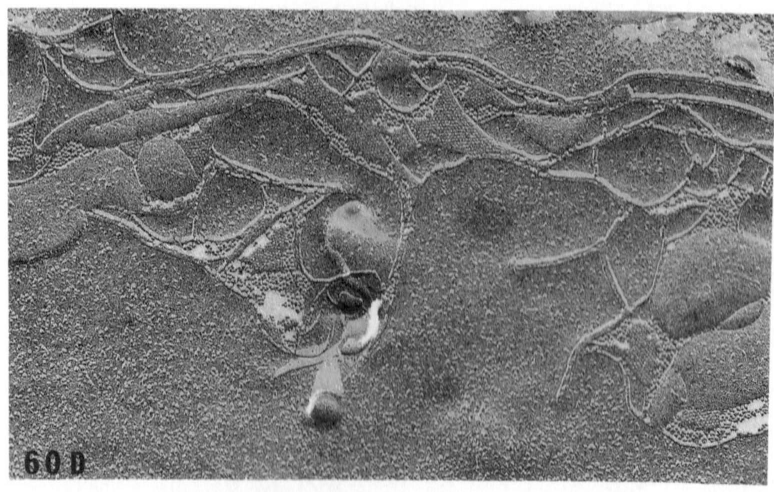

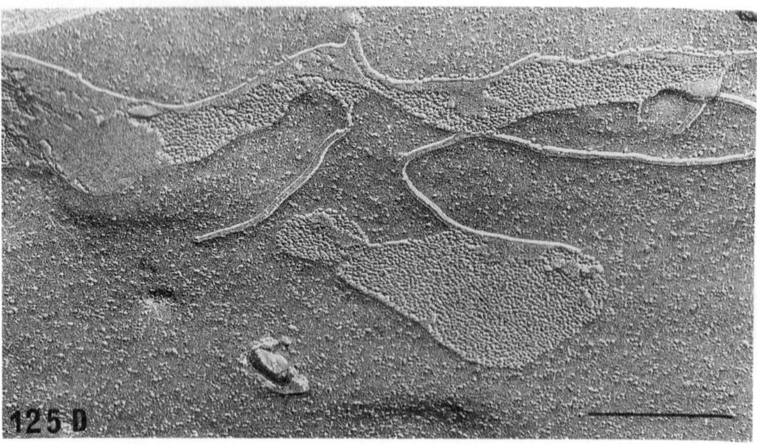

Figure 12.3 Freeze–fracture appearance of tight junctions in choroid plexus of 60 day and 125 day fetal sheep. Upper micrograph shows freeze–fracture replica of a choroid plexus epithelial cell from a 60 day fetal sheep. The fracture has exposed a large area of tight junction towards the apex (CSF surface) of the cell. There are at least four strands running roughly parallel to the apical cell surface. Lower micrograph shows freeze–fracture replica of a choroid plexus epithelial cell from a 125 day fetal sheep. Note the similar number of strands compared with 60 days and that the junctional depth is at least as great at 60 days as at 125 days; bar indicates 0.5 μm. From Møllgård *et al.* (1979)

immature brain (see Figure 12.1). In the adult brain, protein (e.g. horseradish peroxidase), if introduced into the cerebral ventricles, will penetrate freely between the cells of the ependyma to reach the extracellular spaces of the brain (Brightman and Reese, 1969). When such an experiment was performed in an early-stage fetus (sheep: Fossan *et al.*, 1985), it was found that HRP, although taken up by neuroependymal cells, did not penetrate into the extracellular spaces of the brain; however, in late-stage gestation fetal sheep (125 days; term is 150 days from conception) it did. The explanation for this unexpected finding lies in the presence of membrane specializations between adjacent neuroepithelial cells. These are illustrated in thin-section EM in Figure 12.4 and in freeze–fracture in Figure 12.5. It is the freeze-fracture result which shows these to be unusual intercellular junctions. They consist of a single strand of membrane proteins, which is unlike the strands of tight junctions in both orientation and dimensions of membrane particles. Tight junction strands (usually several interlacing strands) are arranged circumferentially towards the apex of a cell, completely sealing off the intercellular space between adjacent cells. These strap junctions consist of a single strand of membrane particles that are larger than those found in tight junctions. In addition, the single strand runs more or less at right angles to the direction one would expect for a tight junction. The strap junctions spiral perpendicularly from the CSF surface of the ventricular zone cells towards the base of these cells within the neuroepithelial layer. The strap junctions have the effect of restricting the width of the intercellular space, as would also be the case for tight junctions. The open pathway for diffusion which is characteristic of the adult is present in late fetal sheep, as is illustrated in Figure 12.4, by which time no strap junctions are present.

The functional effect of these junctions is to complete the isolation of the developing brain from high concentrations of protein in plasma and CSF. The functional significance of this isolation is a matter for speculation (see 'Concluding Speculative Summary', below).

THE PERMEABILITY OF BRAIN BARRIERS TO MACROMOLECULES OF THE SIZE OF PLASMA PROTEIN (>2.5 nm)

The Blood–Brain, Cerebral Endothelial, Barrier

As has been mentioned in the previous section, immunocytochemical evidence (illustrated in Figure 12.1) indicates that plasma proteins do not appear to pass freely from the plasma into brain extracellular space. This has now been studied in a wide range of species: human fetus (Møllgård and Jacobsen, 1984), sheep fetus (Reynolds and Møllgård, 1985), pig fetus (Cavanagh and Møllgård, 1985), rat fetus (Figure 12.1), newborn tammar wallaby (Dziegielewska *et al.*, 1988) and opossum (Saunders *et al.*, 1989). There is additional evidence for the impermeability of the fetal blood–

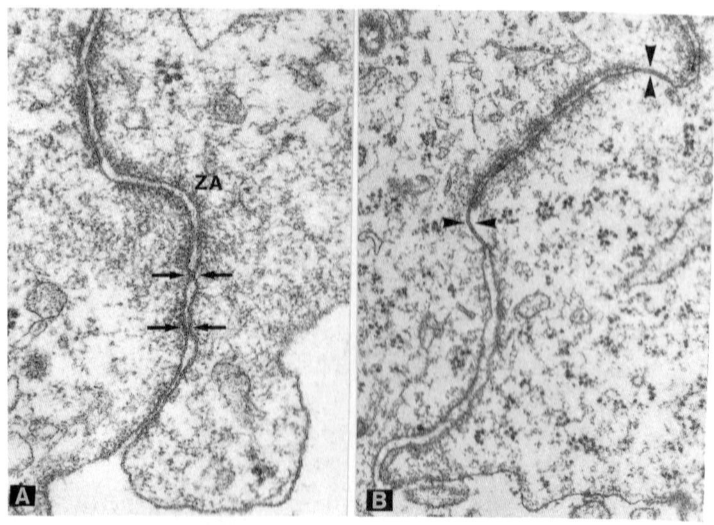

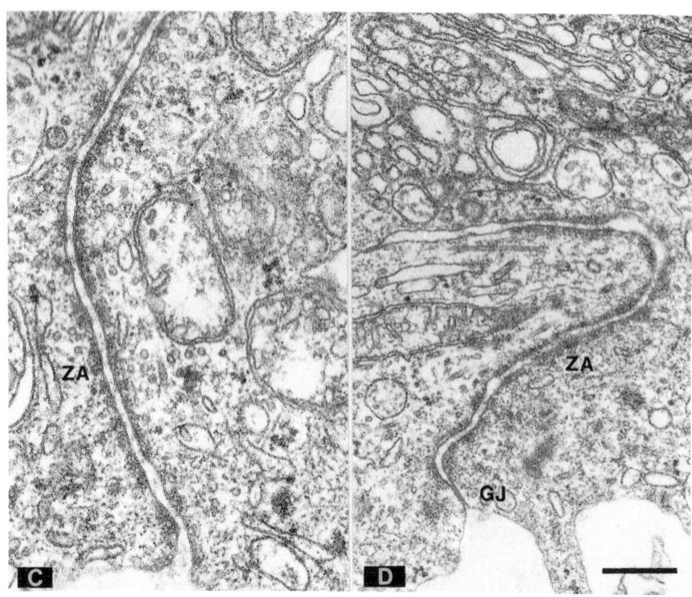

Figure 12.4 Transmission electron micrographs of strap junctions between neuroependymal cells in immature fetal sheep (A, 30 days' gestation; B, 60 days' gestation) and of ependymal cells of mature fetal sheep (C,D, 125 days' gestation). All micrographs are at the same magnification; bar = 0.5 μm. Arrows indicate site of membrane contact of 'strap junction'; arrowheads indicate gap junction. ZA = zonula adherens; GJ = gap junction. From Møllgård *et al.* (1987)

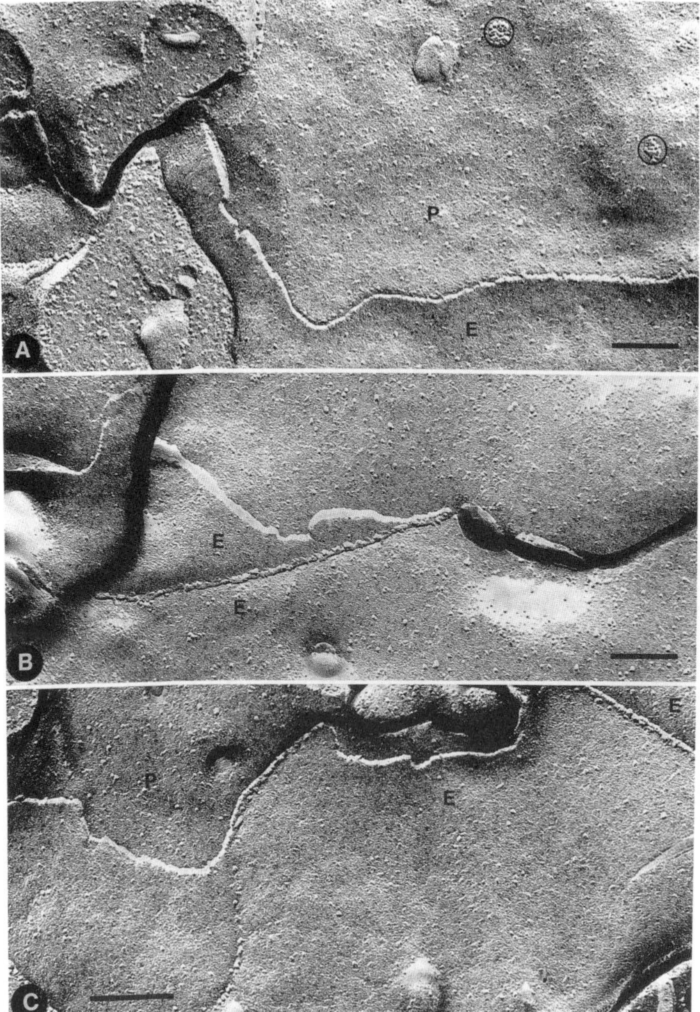

Figure 12.5   P face (A) and E face (B,C) strap junction configurations in 30 day neuro-ependyma. Note the densely packed particulate material in the E face furrows. The encircled particle aggregates may represent small gap junctions but definite identification required complementary particles and pits in the same gap junctional plaque. Bars = 0.2 μm. From Møllgård *et al.* (1987)

brain barrier to protein from experiments using radiolabelled proteins (Olsson *et al.*, 1975) and HRP in moderate amounts (Dziegielewska *et al.*, 1988).

The cerebral endothelial cells of the fetal brain appear to contain some plasma proteins (Figure 12.1; Møllgård *et al.*, 1988b). mRNA for some plasma proteins that are found by immunocytochemistry to be exclusively

Table 12.2 Comparison of freeze–fracture tight junction morphology at 40 and 125 days' gestation in fetal sheep (term is 150 days). Minimum strand number and junctional depth were measured within a standard grid. Mean ± SD; n = number of areas of freeze–fracture replicas counted

Gestational age	Minimum strand no.	Junctional depth	n
40 d	3.64 ± 1.12	0.34 ± 0.17 μm	121
125 d	3.45 ± 1.13	0.28 ± 0.14 μm[a]	49

[a] Significantly different from 40 days. From Møllgård *et al.* (1976).

located in cerebral endothelial cells suggests that these plasma proteins (e.g. α_2-macroglobulin) are synthesized within cerebral endothelial cells. Others may be taken up—e.g. transferrin. It is possible that some of these proteins are transferred to astrocytes, which are also immunocyto-chemically positive for some plasma proteins (Møllgård *et al.*, 1988b). However, there is no evidence of a generalized 'leak' of protein around cerebral vessels in the fetal brain, as would be expected if this barrier were immature.

The Blood–CSF, Choroid Plexus Epithelial, Barrier

Cerebrospinal fluid contains a very high concentration of protein in the fetus, especially early in development. This is illustrated in Figure 12.6 for a number of different species. All are characterized by a high concentration early in brain development. The peak does not appear to be related to the time of birth. Indeed in species that are particularly immature at birth, i.e. marsupials, the peak in CSF protein concentration occurs postnatally. Extensive qualitative and quantitative studies of proteins in fetal and neonatal CSF have been carried out (for review, see Dziegielewska and Saunders, 1988; Dziegielewska *et al.*, 1989a). It seems clear that most, if not all, of the proteins in CSF in the immature brain are immunologically identical with those in plasma. The most comprehensive studies of plasma–CSF exchange of proteins during development have been carried out in the sheep fetus (Dziegielewska *et al.*, 1979, 1980b, 1989b, 1990). Some of the data for these experiments are summarized in Figure 12.7, which illustrates results of experiments carried out in 60 day fetal sheep. At 60 days' gestation the total protein concentration in cisternal CSF is about 350 mg/100 ml (see Figure 12.6), which is about 35% of the peak concentration that occurs at around 30 days' gestation. In Figure 12.7 the results of permeability studies with a variety of marker proteins show the natural steady state for the main proteins in fetal CSF and that this is approached

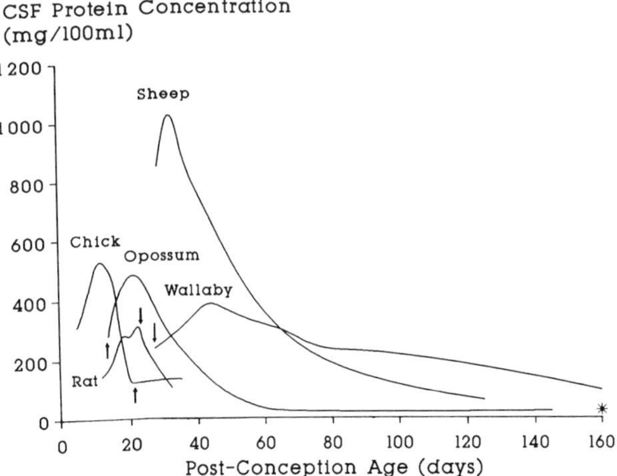

Figure 12.6 Total protein concentration in CSF from cisterna magna of various species during development, estimated by Lowry *et al.* (1951) or Bradford (1976) method. Abscissa: age from conception in days; ordinate: protein concentration in mg/100 ml. * indicates adult value in mammalian species. Birth occurs: in opossum, 14 days (weaned 60 days); chicken and rat, 21 days; wallaby, 28 days (final pouch exit 250 days); sheep, 150 days. From Dziegielewska and Saunders (1990)

by 3–6 h after beginning a slow injection of the marker protein. Although there is some relation between molecular size and the steady-state level for each protein (see also section on the blood–CSF, choroid plexus epithelial, barrier, pp. 135–136), there are obvious differences in the steady states for proteins of similar molecular size. A surprising finding was that iodination of proteins generally reduced their ability to penetrate into CSF and in the case of albumin there appears to be a difference in permeability for different species of albumin (Figure 12.7; see also Dziegielewska *et al.*, 1989b). A similar ability to distinguish between different species of albumin has been found in neonatal rats (Dziegielewska *et al.*, 1989b) and in postnatal *Monodelphis domestica* (grey short-tailed opossum: Knott *et al.*, 1990). Choroid plexus cells from the sheep experiments stain immunocytochemically for both endogenous sheep albumin and the exogenous 'foreign' albumin (unpublished observations), suggesting a possible transcellular route for transfer (cf. Møllgård and Saunders, 1977; Møllgård *et al.*, 1988a). As illustrated in Figure 12.6, the concentration of protein in CSF in the immature brain falls steeply from a peak. Studies later in brain development show that the transfer of proteins from blood to CSF is much less and the specific discrimination between different albumins and different proteins has disappeared (Dziegielewska *et al.*, 1980b; Habgood, 1990); only molecular size discrimination remains (Figure 12.10), apart

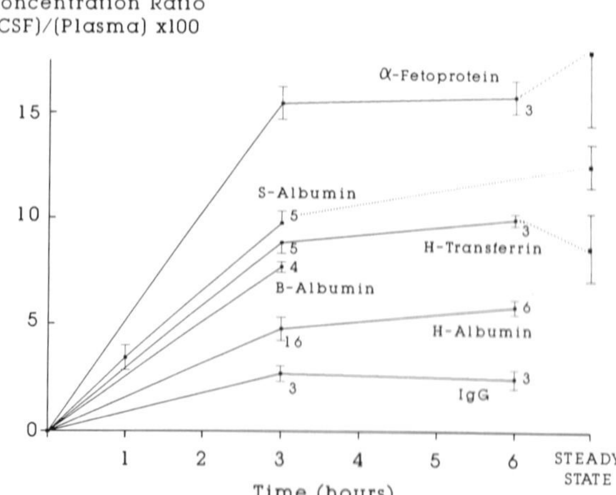

Figure 12.7 Penetration of proteins from plasma into CSF of 60 day fetal sheep. Proteins were injected intravenously and blood was sampled to give an estimate of mean plasma concentration. At times indicated CSF was sampled from cisterna magna. Concentrations of protein in CSF and plasma were estimated by radial immunodiffusion assay. Abscissa: time in hours following i.v. injection. Ordinate: CSF concentration/plasma concentration ×100. Steady state indicates CSF/plasma ratio for naturally occurring sheep proteins. Mean ± SEM. Numbers indicate numbers of experiments. All injected proteins were human, except for S-albumin ([^{125}I]-sheep albumin) and B-albumin (bovine albumin measured using sheep antibovine albumin antiserum). Note (1) that there is some relation between molecular weight and permeability (the largest molecule, IgG, has the smallest ratio and the smallest molecule, AFP, has the largest ratio); (2) proteins of similar size can have significantly different ratios; and (3) albumins from different animal species have different ratios. This suggests that there is a highly selective mechanism that transfers proteins from plasma to CSF in addition to the greater passive permeability illustrated in Figure 12.9. This specific protein transfer mechanism is not present after about 70 days' gestation (4); natural steady states for α-fetoprotein, albumin and transferrin are approached by 3–6 h after injection. From Dziegielewska and Saunders (1990)

from a few proteins such as prealbumin which are synthesized and secreted by the choroid plexus (Thomas *et al.*, 1988).

The presence of well-formed tight junctions between adjacent choroid plexus epithelial cells (see previous section), the intracellular staining of choroid plexus epithelial cells for plasma proteins and the evidence for a high level of apparently species- and protein-specific transfer from blood to CSF in the immature sheep fetus, neonatal rat and neonatal opossum all suggest that there is a developmentally regulated protein transport mechanism in the choroid plexus of the immature brain.

In the adult brain, although the concentration of protein in CSF is very low, there is a gradient of concentration from the lateral ventricles (15 mg/100 ml) through the cisterna magna (25 mg/100 ml) to the lumbar space (45 mg/100 ml: see Davson *et al.*, 1987). In the fetus there is also a gradient, although at a much higher level. Thus, Cavanagh *et al.* (1983), in

60 day gestation fetal sheep, estimated the total protein concentration in lateral ventricular CSF to be about 115 mg/100 ml and in cisternal CSF to be about 460 mg/100 ml; these estimates were based on the sum of the concentrations of the five main proteins in fetal sheep CSF and plasma (α-fetoprotein, α_1-antitrypsin, albumin, fetuin and transferrin). This gives total protein values that are higher than those obtained using the methods of Lowry *et al.* (1951) or Bradford (1976), as in Figure 12.6 (see also Dziegielewska and Saunders, 1988). Very early in brain development (30 days' gestation in the sheep) at the peak of CSF protein concentration (Figure 12.6) little difference was found between lateral ventricular and cisternal CSF (Cavanagh *et al.*, 1983). At this age the brain consists of a sac of CSF surrounded by largely undifferentiated brain cells. It is not clear whether the lack of measured gradient is due to an absence of such a gradient or to the difficulty of obtaining regional samples from such a small brain with relatively large communications between the fluid in different parts of the immature brain.

The CSF–Brain, Neuroepithelial, Barrier

This barrier appears to be a developmental specialization that is only present at early stages of brain development. It has only recently been described (Fossan *et al.*, 1985; Møllgård *et al.*, 1987; and see section on the CSF–brain, neuroepithelial, barrier, pp. 135–136). The permeability of this barrier to proteins has been little studied. HRP in artificial CSF perfused through the ventricular system of immature (60 day gestation) fetal sheep was taken up by neuroepithelial cells but did not penetrate to any extent into the extracellular space (ECS) of the brain (Figure 12.8). In contrast, later in gestation the HRP penetrated freely into brain ECS, as previously described in the adult brain by Brightman and Reese (1969).

Only a few studies involving injection or perfusion of plasma proteins into the CSF spaces have been carried out. These studies, in fetal sheep (Fossan *et al.*, 1985), fetal and neonatal rats (Cavanagh and Warren, 1985) and chick embryos (Moro *et al.*, 1984), are consistent in showing limited uptake of various plasma proteins from CSF into neuroependymal cells that may be both protein-specific and age-dependent. More extensive immunocytochemical studies of plasma proteins in the developing brains of various species show that plasma proteins present in fetal CSF are also found within a proportion of cells lining the cerebral ventricles at some stages of brain development (e.g. human, Møllgård and Jacobsen, 1984; sheep, Reynolds and Møllgård, 1985; pig, Cavanagh and Møllgård, 1985).

THE PERMEABILITY OF BRAIN BARRIERS TO MACROMOLECULES OF THE SIZE OF INULIN (1.3 nm) AND LESS

The Blood–Brain, Cerebral Endothelial, Barrier

Stern (see, e.g., Stern *et al.*, 1929) appears to have been the first to recognize the greater permeability of the developing brain to smaller ('crystalloid') molecules such as sodium ferrocyanide (MW 304). Stern studied the permeability of the fetal and postnatal brain to sodium ferrocyanide in a number of different species, using the Prussian blue reaction to demonstrate the presence (or absence) of sodium ferrocyanide. Stern reported that the immature brain was substantially more permeable to sodium ferrocyanide than is the adult brain.

The first quantitative studies of blood–brain barrier permeability to small-molecular-weight compounds such as inulin (MW 5200, molecular radius 1.3 nm) or sucrose (MW 342, molecular radius 0.51 nm) were carried out by Ferguson and Woodbury (1969). These were pioneering studies, although they suffered from a number of technical limitations, as reviewed elsewhere (Saunders, 1977). The main problems with these experiments were that nothing was known of the physiological state of the neonatal rats used, nor was it determined whether the marker material injected (intraperitoneally) maintained a stable concentration within the plasma. Also, no account was taken of blood contamination of brain samples.

The detailed studies of Habgood (1989, 1990), which are discussed in the next section, indicate that there is indeed a progressive fall in the blood level of radioactive inulin or sucrose which starts several hours after an intraperitoneal injection in nephrectomized neonatal rats. The fall is apparent at about 6 h after injection. This probably accounts for the progressive rise in brain/plasma ratios in Ferguson and Woodbury's experiments for even as long as 15–20 h after injection. Their results are therefore likely to be a substantial overestimate of the permeability of the blood–brain barrier (to which blood contamination of brain samples will have added). However, their general conclusion that the blood–brain barrier to small molecules such as inulin and sucrose is more permeable in immature brains is supported by later experiments. Whether this greater apparent permeability is due to a direct increase in permeability at the endothelial interface between blood and brain or to a reduced sink effect in the immature brain is considered below (pp. 149–150).

More comprehensive studies under well-controlled physiological conditions have been possible in fetal sheep (Dziegielewska *et al.*, 1979). This is because of the relatively larger size of the fetus in this species and the anatomical structure of the placenta. In the sheep this consists of a large number of individual placentae, all connected together to form the two

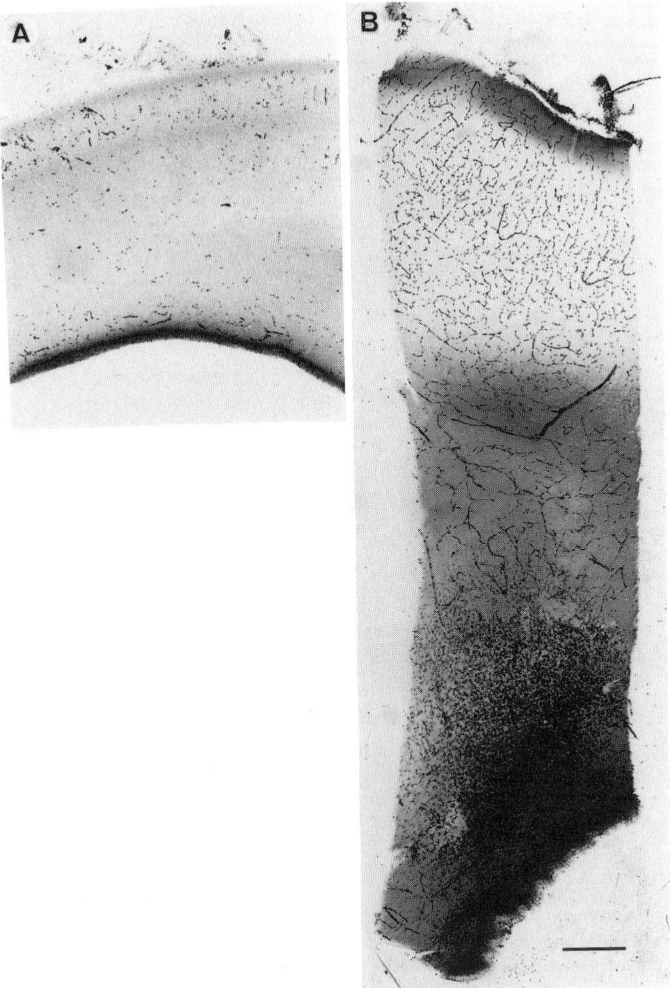

Figure 12.8 Low-magnification light micrographs of coronal sections of the telencephalic wall of a 60 day (A) and 125 day (B) sheep fetus after 5 h of ventriculocisternal perfusion with HRP in artificial CSF. The pial layer at the surface is at the top and the lateral ventricle at the bottom. Peroxidase can be localized 3 mm away from the CSF–brain interface in the brain of the 125 day fetus, whereas the peroxidase activity was confined to the ventricular zone in the 60 day fetus. Note the filling of the subpial layer in the 125 day brain. Peroxidase activity within blood vessels was associated with erythrocytes. The sections are about 80 μm thick and of the same magnification. Bar indicates 0.2 mm. From Fossan *et al.* (1985)

arteries and two veins of the umbilical cord (Barcroft, 1938). It is possible, for example, in 60 day gestation sheep (term is 150 days) to cannulate an artery and vein of a single placenta (of which there are up to 100), thus only interfering with the placental circulation to a minimal extent. Using this preparation, a wide range of molecular markers of different sizes has been

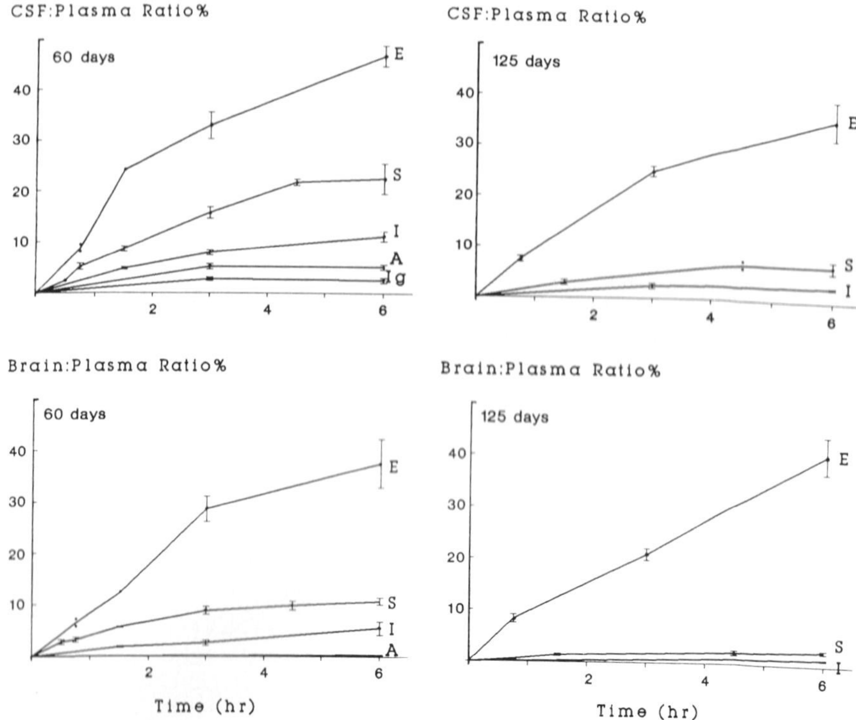

Figure 12.9 Permeability of blood–brain and blood–CSF barriers in anaesthetized fetal sheep at 60 and 125 days' gestation (term is 150 days from conception). Fetuses were given i.v. infusions or intermittent injections to maintain approximately constant blood levels of a wide range of molecular size markers: erythritol [E, mol. radius 0.35 nm]; sucrose [S, 0.51 nm]; inulin [I, 1.3 nm]; human albumin [A, 3.5 nm]; human IgG [Ig, 5.3 nm]. Experiments were run for 20 min to 6 h. Three to six experiments were performed for each time point except where individual points are shown. Bars are ± standard error of mean. Results are expressed as the concentration ratio CSF:plasma ×100 or brain:plasma ×100. The ratios for erythritol at both ages and for CSF and brain are similar, probably because erythritol is small enough to pass through cell membranes. The ratios for brain are smaller than for CSF at each age, because the markers are distributed in brain within the extracellular space. Note the decline in ratios with age, which is thought to reflect a decline in passive permeability with age. The values for human albumin in 60 day fetal brain are not significantly above background blood contamination of the brain samples—i.e. the blood–brain barrier to protein is well-formed even this early in gestation. From Dziegielewska and Saunders (1990)

studied, as is illustrated in Figure 12.9. This illustrates the change in blood–brain barrier permeability to molecules in the size range erythritol (M_r, 0.35 nm) to inulin (M_r, 1.3 nm). Albumin and IgG did not penetrate into brain to any greater extent than could be accounted for by blood contamination of brain samples. These results show the marked decline in apparent permeability to small-molecular-weight compounds at the blood–brain interface which occurs during brain development.

The Blood–CSF, Choroid Plexus Epithelial, Barrier

Following on from the qualitative studies of Stern *et al.* (1929) and the early quantitative studies of Ferguson and Woodbury (1969), detailed experiments using a wide range of size of molecular markers have been undertaken in fetal sheep (Dziegielewska *et al.*, 1979) and neonatal rats (Habgood, 1989, 1990). The experiments of Habgood (1989, 1990) in neonatal rats are particularly significant in emphasizing the importance of obtaining a stable blood level of marker if reliable estimates of the true CSF/plasma steady-state ratio are to be obtained. By using a standardized intraperitoneal injection technique in 2-day-old nephrectomized rats, Habgood (1989) has shown that very similar levels of $[^{14}C]$-sucrose or $[^{14}C]$-inulin can be achieved in the plasma of neonates from the same litter. It was therefore possible to carry out time-dependent permeability studies using a litter of 2 day rats as the experimental model and sampling pairs of rats (CSF and plasma) at hourly or two-hourly intervals. These experiments showed that a plateau of concentration of radioactive label was achieved between about 2 and 5 h post injection, after which the plasma level of radioactivity began to fall. By 4–6 h the level of radioactivity in CSF was stable. Thus, the CSF/plasma ratios at 4–6 h for inulin (13% ± 0.3 SEM) and sucrose (32% ± 0.9) were representative of the steady-state CSF/plasma ratios for these molecules. In the earlier experiments of Ferguson and Woodbury (1969) similar CSF/plasma ratios were achieved at 6 h, but the CSF/plasma ratios continued to increase for up to 24 h, reaching 80% for sucrose and 30% for inulin. It is clear from Habgood's experiments that this is likely to have been due to a progressive decline in plasma levels of radioactive marker. The fetal sheep blood–CSF permeability experiments are from the same series as the blood–brain barrier studies described in the previous section and are also illustrated in Figure 12.9. As for the blood–brain barrier, these results show a marked decline in the apparent permeability of the blood–CSF barrier with age. Because this interface is also permeable to protein, it was possible to study penetration into CSF of a much wider range of molecular size.

Felgenhauer (1974), from his studies of proteins in adult human CSF, has suggested that, with a few exceptions, the concentration ratio of individual plasma protein betwen CSF and plasma is a reflection of the molecular size and diffusion coefficient of the protein. Thus, a log-linear plot of concentration ratio (Felgenhauer used serum/CSF rather than the more usual CSF/plasma) gives a straight line over a wide range of molecular size. Such a plot (but as CSF/plasma) is illustrated in Figure 12.10 for his human data and compared with the data from sheep at different ages. For sucrose, inulin and the *human* protein studies in 60 day fetal sheep the data fall on a straight line approximately parallel to the plot for the human protein data. The sheep data from 125 days' gestation fall on

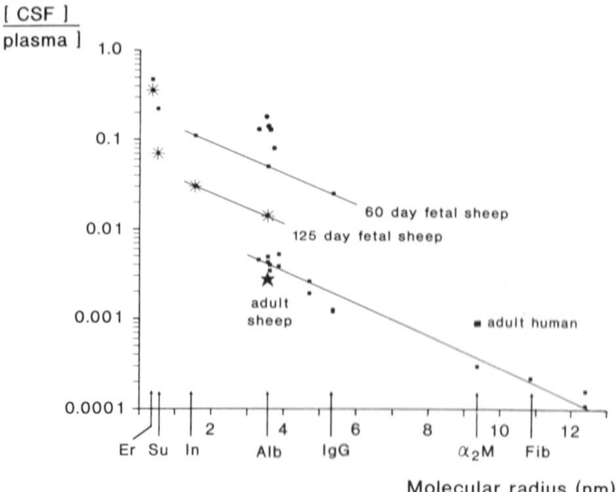

Figure 12.10 Inverted Felgenhauer (1974) plot of CSF/plasma ratio (log scale) against molecular radius (nm) for compounds of different molecular sizes. Er = erythritol; Su = sucrose; In = inulin; Alb = albumin; IgG = immunoglobulin G; α_2M = α_2-macroglobulin; Fib = fibrinogen. Sheep data from Figure 12.9. Points above 60 day line are sheep protein data from Figure 12.7 and unpublished data. Human data from Felgenhauer (1974). Black star indicates adult sheep albumin. Note parallel log linear lines with decreasing CSF/plasma ratios as age increases. This relationship suggests unrestricted diffusion through very large pores, with a reduction in the number of pores rather than their size during development. Values for sucrose and especially for erythritol are above the line, suggesting an additional population of small pores. Erythritol is probably small enough to pass through cell membranes. Note protein points above 60 day fetal sheep line. These are values for the five main proteins in CSF and plasma of fetal sheep: α-fetoprotein, α_1-antitrypsin, albumin, fetuin and transferrin. Their CSF/plasma ratios are all higher than can be accounted for by diffusion and may be explained by a specific transfer of proteins from blood to CSF, probably via the choroid plexus (especially IVth ventricular); cf. data in Figure 12.7

an intermediate but parallel line. Data for sucrose at 60 and 125 days lie somewhat above the line, which indicates a proportionately greater permeability to such small molecules. Permeability to the still smaller erythritol (M_r, 0.35 nm) is disproportionately greater than for sucrose; unlike permeability to all of the larger molecules, permeability to erythritol is similar at both 60 and 125 days. It may be that erythritol is sufficiently small to pass directly through cell membranes of cerebral endothelia and choroid plexus epithelia. Data for *sheep* endogenous proteins are included for 60 days' gestation. All fall well above the line, indicating that there is more present in CSF than can be accounted for by a mechanism based solely on molecular size and diffusion coefficient. This plot appears to provide a convenient way of distinguishing between diffusion-dependent blood–CSF exchange and more specific (protein carrier?) mechanisms. Evidence for protein-specific transfer across the immature choroid plexus was reviewed in the section on the blood–CSF, choroid plexus epithelial, barrier (pp. 140–142). There is also evidence for synthesis and secretion of some plasma proteins in immature (Dziegielewska *et al.*, 1984; Saunders, 1984; Thomas *et al.*, 1988) and even adult (Thomas *et al.*, 1988, 1989) choroid plexus which may account for the disproportionately high CSF/plasma ratio reported for a few proteins such as prealbumin. The parallel decline in the passive permeability lines between 60 days and the adult suggests that the decline in apparent permeability is due to a rather generalized mechanism that is affecting a wide range of molecular size proportionately. The possible effect of the increase in CSF secretion rate (sink effect), occurring with age, on the apparent blood–CSF and blood–brain permeability of molecules introduced into the plasma will be considered in the next section. However, this is unlikely to account for the changes illustrated in Figure 12.10, since an increase in CSF secretion rate would be expected to have a greater effect on the steady-state ratio of larger molecules (see Dziegielewska *et al.*, 1979, p. 211); this would be expected to change the slope of the lines, with age, in Figure 12.10. It seems more likely that the age-related change in apparent permeability to molecules in the size range sucrose to inulin in brain and sucrose to IgG in CSF can be accounted for by a reduction in the surface area for exchange. If the site for exchange is in the form of 'pores', then these are necessarily large and unchanging in diameter with age, since the data of Figure 12.10 suggest that no differential decrease in permeability for larger molecules occurs with increasing age.

CSF-sink Effect in the Developing Brain

The absolute rate of CSF secretion in the immature brain is undoubtedly much less than in the adult. Thus, in 3-day-old rats Woodbury *et al.* (1974) estimated a rate of 0.2 μl/min. Values for sheep of different ages are shown in Table 12.3. Several authors (e.g. Woodbury *et al.*, 1974; Amtorp, 1976) have suggested that the apparently greater permeability of the developing brain might be due to a lower sink effect. Unfortunately, it is not entirely clear on what basis brains of different ages (which are enormously different in size: see Table 12.3) can be compared. Johanson and Woodbury (1974) expressed their results for CSF secretion in neonatal rats as volume per minute per unit weight of choroid plexus. This showed that the secretory capacity of the choroid plexus increases considerably with age, but does not tell us how this capacity relates to the increasing size of the brain. An alternative approach is shown in Table 12.3, in which the secretion rate is expressed as turnover (i.e. secretion rate as a percentage of the total volume). This is remarkably similar at the three different ages studied. Davson *et al.* (1987) have indicated a comparable striking similarity in the turnover of CSF in adults of a wide variety of species with very different

Table 12.3   CSF secretion in fetal and adult sheep: estimates of CSF turnover and 'sink effect' related to brain size at different ages. Data on secretion rate at 60 days are from Fossan *et al.* (1985). Rest of data are from Evans *et al.* (1974). The latter authors obtained a much higher estimate of CSF secretion rate in 60 day fetuses, probably because the experiments were not long enough for a steady state to have been reached. Note large increase in secretion rate with age, but little change in turnover

Age (days from conception)	*n*	*Brain weight* (g)	*CSF secretion* (μl/min)	*CSF vol.* (ml)	*Turnover* (%/min)	*'sink'* (%/min per g *brain weight*)
60	5	1.91 ± 0.18	2.8 ± 0.38	0.45 ± 0.04	0.62	0.32
122	15	36.7 ± 1.0	62.5 ± 7.9	7.12 ± 1.3	0.88	0.02
Adult	12	78.4 ± 2.6	118.4 ± 12.8	14.2 ± 2.4	0.83	0.01

brain sizes. If the turnover of CSF is related to the size of the brain for which it is acting as a sink, the surprising result is obtained that on this basis CSF secretion is relatively less in older brains (Table 12.3). This seems intrinsically unlikely, as would be the similar conclusion for different adult species if the data of Davson *et al.* (1987) were calculated on the basis of brain weight. However, even if only the turnover data (Table 12.3) are taken as representative of the sink effect in brains of different ages (and sizes), then it is clear that the sink effect is not contributing to the decline in apparent permeability with age which is illustrated in Figures 12.9 and 12.10. This reinforces the conclusion drawn in the previous section that the decline in apparent permeability to macromolecules that occurs in the developing brain is likely to be due to a genuine change in permeability at the blood–CSF and blood–brain interfaces.

CONCLUDING SPECULATIVE SUMMARY

The whole of the research described in this chapter was initiated as a result of a fortunate association with Hugh Davson when we were colleagues at University College London. The outcome in terms of a description of the barriers to macromolecules in the developing brain can be summarized rather simply: the original and fundamental barrier to protein, which is still widely believed to be immature in the developing brain, is in fact well-formed, at both the blood–brain and blood–CSF interfaces. However, at the latter site the barrier (tight junctions) is by-passed by a transcellular route across the choroid plexus. On the other hand, a barrier which had not been previously considered, because it does not exist in the adult brain, does occur in the immature brain—namely, a barrier to protein at the CSF–brain interface (strap junctions). The overall effect of these barrier mechanisms to protein in the developing brain is that there is a high concentration of protein in CSF but neither this protein nor that in cerebral blood vessels is able to penetrate directly into the extracellular space of the brain. Thus, even the immature brain appears to be 'protected' from extracellular protein except at the level of the dividing neuroepithelial cells that line the cerebral ventricles and give rise, postmitotically, to the different layers and regions of the developing brain.

It may be that the exclusion of plasma proteins from the extracellular space of the developing brain is important for the normal differentiation of neurons. There is some evidence from tissue culture studies that neurons, as opposed to satellite cells, do less well in culture in the presence of the higher concentrations of fetal calf serum that are often included in culture media (Coon and Sinback, 1982). Some neuron populations in the developing brain contain specific plasma glycoproteins (e.g. fetuin, transferrin), which they probably synthesize *in situ* (Møllgård *et al.*, 1988a). Bruckenstein and Higgins (1988) have demonstrated stimulatory effects of fetal calf serum and fractions of such serum on dendritic growth of sympathetic neurons in culture. It is possible that proteins, such as fetuin and transferrin, act as trophic or identification signals between different neuronal populations. Such a mechanism would be ineffective if the extracellular space of brain were flooded with plasma proteins entering via an immature blood–brain barrier. It is perhaps significant that the strap junctions forming the CSF–brain barrier disappear in the sheep fetus after about 60 days' gestation, a time when neuronal mitosis is largely over and glial cell proliferation takes over (Åström, 1969).

The high concentration of plasma proteins in fetal CSF, which seems to be achieved partly by a greater passive permeability and partly by a specific protein transfer system, may be important for some aspects of the development of the neuroepithelium which lines the cerebral ventricles of the developing brain and with which these proteins are in direct contact. Neuroepithelial cells take up certain proteins from fetal CSF. The specificity and tissue dependence of this process has not been adequately determined, but these proteins may contribute to mitosis and the decision to migrate of subpopulations of neuroepithelial cells. Better understanding of this process would not only lead to improved understanding of the normal process of brain development, but also might lead to methods for *in vitro* culture of neuroepithclial (ventricular zone) cells with continued mitosis, which might be valuable in attempts to use fetal cell grafts to repair the brain damage of such conditions as Parkinson's disease (e.g. Moore, 1987).

ACKNOWLEDGEMENTS

I should like to thank Hugh Davson and all my many friends and colleagues at University College London, Southampton and elsewhere for their advice, support and collaboration over many years; without the enormous contribution and support of Katarzyna Dziegielewska most of this work would not have been possible. I should also like to thank Kjeld Møllgård for supplying original micrographs and for helpful discussion of the manuscript, in addition to his considerable contribution to our joint work over many years. Also, my thanks to my secretary, Lynn Ford, for typing the manuscript and to Mark Habgood and Gabrielle Turner for preparing the figures.

The studies described in this chapter were carried out over the period 1969 to 1989 with the aid of grants from Action Research, AFRC, MRC, NATO, Nuffield Foundation and The Wellcome Trust, for whose support I am most grateful.

REFERENCES

Amtorp, O. (1976). Transfer of I^{125}-albumin from blood into brain and cerebrospinal fluid in newborn and juvenile rats. *Acta Physiol. Scand.*, **96**, 399–406

Åström, K.-E. (1967). On the early development of the isocortex in fetal sheep. In Bernhard, C. G. and Schade, J. P. (Eds), *Progress in Brain Research*, Vol. 26: *Developmental Neurology*. Elsevier, Amsterdam, London, New York, pp. 1–59

Barcroft, J. (1938). *The Brain and Its Environment*. Yale University Press, New Haven

Behnsen, G. (1927). Uber die Farbstoffspeicherung im Zentralnervensystem der Weissen Maus in verschiedenen Alterszustanden. *Zeit. Zell. Microsk. Anat.*, **4**, 515–572

Bradford, M. M. (1976). A rapid sensitive method for the quantitation of microgram quantities of protein utilizing the principle of protein–dye binding. *Anat. Biochem.*, **72**, 248–254

Brightman, M. W. and Reese, T. S. (1969). Junctions between intimately apposed cell membranes in the vertebrate brain. *J. Cell Biol.*, **40**, 648–677

Bruckenstein, D. A. and Higgins, D. (1988). Morphological differentiation of embryonic rat sympathetic neurons in tissue culture. II. Serum promotes dendritic growth. *Dev. Biol.*, **128**, 337–348

Cavanagh, M. E., Cornelis, M. E. P., Dziegielewska, K. M., Evans, C. A. N., Lorscheider, F. L., Møllgård, K., Reynolds, M. L. and Saunders, N. R. (1983). Comparison of proteins in csf of lateral and IVth ventricles during early development of fetal sheep. *Dev. Brain Res.*, **11**, 159–167

Cavanagh, M. E. and Møllgård, K. (1985). An immunocytochemical study of the distribution of some plasma proteins within the developing forebrain of the pig with special reference to the neocortex. *Dev. Brain Res.*, **17**, 183–194

Cavanagh, M. E. and Warren, A. (1985). The distribution of native albumin and foreign albumin injected into lateral ventricles of prenatal and neonatal rat forebrains. *Anat. Embryol.*, **172**, 345–351

Coon, H. G. and Sinback, C. N. (1982). In Sirbasku, D. A. *et al.* (Eds), *Growth of Cells in Hormonally Defined Media*, Book B. Cold Spring Harbor, New York, pp. 1007–1016

Davson, H., Welch, K. and Segal, M. B. (1987). *The Physiology and Pathophysiology of the Cerebrospinal Fluid*. Churchill Livingstone, Edinburgh

Dziegielewska, K. M., Evans, C. A. N., Fossan, G., Lorscheider, F., Malinowska, D. H.,

Møllgård, K., Saunders, N. R. and Wilkinson, S. (1980a). Proteins in cerebrospinal fluid and plasma of fetal sheep during development. *J. Physiol. (London)*, **300**, 441–455

Dziegielewska, K. M., Evans, C. A. N., Malinowska, D. H., Møllgård, K., Reynolds, J. M., Reynolds, M. L. and Saunders, N. R. (1979). Studies of the development of brain carrier systems to lipid insoluble molecules in fetal sheep. *J. Physiol. (London)*, **292**, 207–231

Dziegielewska, K. M., Evans, C. A. N., Malinowska, D. H., Møllgård, K., Reynolds, M. L. and Saunders, N. R. (1980b). Blood–cerebrospinal fluid transfer of plasma proteins during fetal development in the sheep. *J. Physiol. (London)*, **300**, 457–465

Dziegielewska, K. M., Evans, C. A. N., New, H., Reynolds, M. L. and Saunders, N. R. (1984). Synthesis of plasma proteins by rat fetal brain and choroid plexus. *Int. J. Devl. Neurosci.*, **2**, 215–222

Dziegielewska, K. M., Habgood, M., Jones, S. E., Reader, M. and Saunders, N. R. (1989a). Proteins in cerebrospinal fluid and plasma of postnatal *Monodelphis domestica* (grey short-tailed opossum). *Comp. Biochem. Physiol.*, **92B**, 569–576

Dziegielewska, K. M., Habgood, M. D., Møllgård, K. and Saunders, N. R. (1990). Transfer of plasma proteins from blood into different cerebrospinal fluid compartments in the immature fetal sheep. In preparation

Dziegielewska, K. M., Habgood, M. and Saunders, N. R. (1989b). Species specific blood–CSF transfer of albumin in anaesthetized immature fetal sheep. *J. Physiol. (London)*, **415**, 99P

Dziegielewska, K. M., Hinds, L. A., Møllgård, K., Reynolds, M. L. and Saunders, N. R. (1988). Blood–brain, blood–cerebrospinal fluid and cerebrospinal fluid–brain barriers in a marsupial (*Macropus eugenii*) during development. *J. Physiol. (London)*, **403**, 367–388

Dziegielewska, K. M. and Saunders, N. R. (1988). The development of the blood–brain barrier: proteins in fetal and neonatal CSF, their nature and origins. In Meisami, E. and Timiras, P. J. (Eds), *Handbook of Human Growth and Biological Development*, Vol. 1, Part A. CRC Press, Boca Raton, Forida, pp. 169–191

Dziegielewska, K. M. and Saunders, N. R. (1990). The internal environment of the developing brain. In Mednick, S. A. (Ed.), *Development and Neuropathology of Schizophrenia*, NATO Advanced Research Workshop. Plenum Press, New York (in press)

Ehrlich, P. (1885). In *Das Sauerstoff-Bedurfniss des Organismus*. Eine farbenanalytische Studie. Hirschwald, Berlin, pp. 69–72

Evans, C. A. N., Reynolds, J. M., Reynolds, M. L., Saunders, N. R. and Segal, M. B. (1974). The development of a blood–brain barrier mechanism in foetal sheep. *J. Physiol. (London)*, **238**, 371–386

Felgenhauer, K. (1974). Protein size and cerebrospinal fluid composition. *Klin. Wschr.*, **52**, 1158–1164

Ferguson, R. K. and Woodbury, D. M. (1969). Penetration of ^{14}C-inulin and ^{14}C-sucrose into brain, cerebrospinal fluid and skeletal muscle of developing rats. *Exp. Brain Res.*, **7**, 181–194

Fossan, G., Cavanagh, M. E., Evans, C. A. N., Malinowska, D. H., Møllgård, K., Reynolds, M. L. and Saunders, N. R. (1985). CSF–brain permeability in the immature sheep fetus: a CSF–brain barrier. *Dev. Brain Res.*, **18**, 113–124

Ganong, W. F. (1989). *Review of Medical Physiology*, 14th edn. Prentice Hall, London, pp. 452, 518

Grazer, F. M. and Clemente, C. D. (1957). Developing blood–brain barrier to trypan blue. *Proc. Soc. Exp. Biol. Med.*, **94**, 758–760

Habgood, M. D. (1989). Blood–CSF permeability in very immature rats. *J. Physiol. (London)*, **417**, 31P

Habgood, M. D. (1990). *Barriers in the Developing Brain*. PhD thesis, University of Southampton

Jacobsen, M., Clausen, P. P., Jacobsen, G. K., Saunders, N. R. and Møllgård, K. (1982a). Intracellular plasma proteins in human fetal choroid plexus during development. I. Developmental stages in relation to the number of epithelial cells which contain albumin in telencephalic, diencephalic and myelencephalic choroid plexus. *Dev. Brain Res.*, **3**, 239–250

Jacobsen, M., Jacobsen, G. K., Clausen, P. P., Saunders, N. R. and Møllgård, K. (1982b). Intracellular plasma proteins in human fetal choroid plexus during development. II. The distribution of prealbumin, albumin, alpha-fetoprotein, transferrin, IgG, IgA, IgM and alpha$_1$-antitrypsin. *Dev. Brain Res.*, **3**, 251–262

Johanson, C. E. and Woodbury, D. M. (1974). Changes in CSF flow and extracellular space in the developing rat. In Vernadakis, A. and Weiner, N. (Eds), *Drugs and the Developing Brain*. Plenum Press, New York, pp. 281–287

Knott, G., Habgood, M. D. and Dziegielewska, K. M. (1990). Blood–CSF albumin transfer in the neonates of a species of opossum. In preparation

Lowry, O. H., Rosebrough, N. J., Farr, A. L. and Randall, R. J. (1951). Protein measurement with the folin phenol reagent. *J. Biol. Chem.*, **193**, 265–275

Møllgård, K., Balslev, Y., Lauritzen, B. and Saunders, N. R. (1987). Cell junctions and membrane specializations in the ventricular zone (germinal matrix) of the developing sheep brain: a CSF–brain barrier. *J. Neurocytol.*, **16**, 433–444

Møllgård, K., Balslev, Y. and Saunders, N. R. (1988a). Structural aspects of the blood–brain and blood–CSF barriers with respect to endogenous proteins. In Rakić, L., Begley, D. J., Davson, H. and Zloković, B. V. (Eds), *Peptide and Amino Acid Transport Mechanisms in the Central Nervous System*. Macmillan Press, London, pp. 93–101

Møllgård, K., Dziegielewska, K. M., Saunders, N. R., Zakut, H. and Soreq, H. (1988b). Synthesis and localization of plasma proteins in the developing human brain. Integrity of the fetal blood–brain barrier to endogenous proteins of hepatic origin. *Dev. Biol.*, **128**, 207–221

Møllgård, K. and Jacobsen, M. (1984). Immunohistochemical identification of some plasma proteins in human embryonic and fetal forebrain with particular reference to the development of the neocortex. *Dev. Brain Res.*, **13**, 49–63

Møllgård, K., Lauritzen, B. and Saunders, N. R. (1979). Double replica technique applied to choroid plexus from early foetal sheep: completeness and complexity of tight junctions. *J. Neurocytol.*, **8**, 139–149

Møllgård, K. Malinowska, D. H. and Saunders, N. R. (1976). Lack of correlation between tight junction morphology and permeability properties in developing choroid plexus. *Nature*, **264**, 293–294

Møllgård, K. and Saunders, N. R. (1975). Complex tight junctions of epithelial and of endothelial cells in early foetal brain. *J. Neurocytol.*, **4**, 453–468

Møllgård, K. and Saunders, N. R. (1977). A possible transepithelial pathway via endoplasmic reticulum in foetal sheep choroid plexus. *Proc. Roy. Soc. London*, **B199**, 321–326

Møllgård, K. and Saunders, N. R. (1986). The development of the human blood–brain and blood–CSF barriers. *Neuropathol. Appl. Neurobiol.*, **12**, 337–358

Moore, R. Y. (1987). Parkinson's disease—a new therapy? *New Engl. J. Med.*, **316**, 892–893

Moro, R., Fielitz, N., Esteves, A., Grunberg, J. and Uriel, J. (1984). *In vivo* uptake of heterologous alphafetoprotein and serum albumin by ependymal cells of developing chick embryos. *Int. J. Dev. Neurosci.*, **2**, 143–148

Olsson, Y., Klatzo, I., Sourander, P. and Steinwall, O. (1968). Blood–brain barrier to albumin in embryonic, newborn and adult rats. *Acta Neuropathol.*, **10**, 117–122

Penta, P. (1932). Sulla colorazione vitale del sistema nervosa centrale negli animali neonati. *Riv. Neurol.*, **5**, 62–80

Reynolds, M. L. and Møllgård, K. (1985). The distribution of plasma proteins in the neocortex and early allocortex of the developing sheep brain. *Anat. Embryol*, **171**, 41–60

Risau, W., Hallman, R. and Albrecht, U. (1986). Differentiation-dependent expression of proteins in brain endothelium during development of the blood–brain barrier. *Dev. Biol.*, **117**, 537–545

Saunders, N. R. (1977). Ontogeny of the blood–brain barrier. *Exp. Eye Res.*, **25** (Suppl.), 523–550

Saunders, N. R. (1984). Plasma proteins and fetal brain development. In Duprat, A.-M., Kato, A. C. and Weber, M. (Eds), *The Role of Cell Interactions in Early Neurogenesis*. NATO Advanced Study Institute, Plenum Press, New York, pp. 191–199

Saunders, N. R., Adam, E., Reader, M. and Møllgård, K. (1989). *Monodelphis domestica* (grey short-tailed opossum): An accessible model for studies of early neocortical development. *Anat. Embryol.*, **180**, 227–236

Saunders, N. R. and Møllgård, K. (1984). Development of the blood–brain barrier. *J. Dev. Physiol.*, **6**, 45–47

Stern, L. (1934). A propos de la methode d'investigation du fonctionnement de la barrière hemato–encephalique. *C. R. Soc. Biol.*, **115**, 1059–1061

Stern, L. and Peyrot, R. (1927). Le fonctionnement de la barrière hemato–encephalique aux divers stades de developpement chez diverses especes animales. *C. R. Soc. Biol.*, **96**, 1124–1126

Stern, L. and Rapoport, J.-L. (1928). Les rapports entre l'augmentation de la permeabilite de la barrière hemato–encephalique et les alterations de son substratum morphologique. *C. R. Soc. Biol.*, **98**, 1515–1517

Stern, L., Rapoport, J.-L. and Lokschina, E.-S. (1929). Le fonctionnement de la barrière hemato–encephalique chez les nouveau-nes. *C. R. Soc. Biol.*, **100**, 231–233

Stewart, P. A. and Hayakawa, E. M. (1987). Interendothelial junctional changes underlie the developmental tightening of the blood–brain barrier. *Dev. Brain Res.*, **32**, 271–281

Tauc, M., Vignon, X. and Bouchaud, C. (1984). Evidence for the effectiveness of the blood–csf barrier in the fetal rat choroid plexus. A freeze–fracture and peroxidase diffusion study. *Tiss. Cell*, **16**, 65–74

Thomas, T., Power, B., Hudson, P., Schreiber, G. and Dziadek, M. (1988). The expression of transthyretin mRNA in the developing rat brain. *Dev. Biol.*, **128**, 415–427

Thomas, T., Schreiber, G. and Jaworowski, A. (1989). Developmental patterns of gene expression of secreted proteins in brain and choroid plexus. *Dev. Biol.*, **134**, 38–47

Tschirgi, R. D. (1950). Protein complexes and the impermeability of the blood–brain barrier to dyes. *Am. J. Physiol.*, **163**, 756

Vorbrodt, A. W., Lossinsky, A. S. and Wisniewski, H. M. (1986). Localization of alkaline phosphatase activity in endothelia of developing and mature mouse blood–brain barrier. *Dev. Neurosci.*, **8**, 1–13

Wakai, S. and Hirokawa, N. (1981). Development of blood–cerebrospinal fluid barrier to horseradish peroxidase in the avian choroidal epithelium. *Cell Tiss. Res.*, **214**, 271–278

Wislocki, G. B. (1920). Environmental studies on fetal absorption. I. The vitally stained fetus. *Contrib. Embryol.*, **51**, 45–52

Woodbury, D. M., Johanson, C. and Brønsted, H. (1974). Maturation of the blood–brain and blood–cerebrospinal fluid barriers and transport systems. In Zimmerman, E. and George, R. (Eds), *Narcotics and the Hypothalamus*. Raven Press, New York, pp. 225–247

13
Pathophysiology of Communicating Hydrocephalus: Information Provided by the New Imaging Modalities

A. Everette James, Jr., Christine H. Lorenz, James A. McKanna,
Jeff L. Creasy, C. Leon Partain, Ernst-Peter Strecker,
William Bradley, Jr.

INTRODUCTION

This chapter addresses the physiology of CSF with particular emphasis upon the dynamic process of communicating hydrocephalus. Utilizing traditional morphological and physiological techniques of histological and ultrastructural analysis (Price *et al.*, 1976; James *et al.*, 1980; Diggs *et al.*, 1986) correlated with autoradiography (Strecker *et al.*, 1973, 1974), radioactive transfer measurements (James *et al.*, 1970, 1972) and cisternography, we have documented certain associated structural and functional abnormalities in communicating hydrocephalus as well as the compensatory and repair mechanisms. Recently we have employed the modality of magnetic resonance imaging (Partain *et al.*, 1988b,c) and its dynamic capabilities (Price *et al.*, 1987) to further the understanding of CSF physiology as it relates to communicating hydrocephalus (Bradley *et al.*, 1986, 1989; Davson *et al.*, 1987).

Communicating hydrocephalus results in a relative imbalance of production of CSF fluid and its absorption. Absorption is most often compromised by closure of pathways of CSF flow and drainage (Deland *et al.*, 1972). Early in the development of communicating hydrocephalus, the CSF pressure is elevated (Vessal *et al.*, 1974), but as the cerebral ventricles enlarge, the pressure and ventricular entry decrease to the normal range, except for transient episodes of pressure elevation.

The transfer of labelled substances from CSF (Davson *et al.*, 1962) into the blood is also compromised (Strecker *et al.*, 1977), despite the fact that CSF production continues at a normal rate (James *et al.*, 1973). With the cerebral ventricular enlargement at the expense of the neuropil, a series of morphological changes have been observed which suggest an attempt to respond to the physiology. These phenomena remained underinvestigated, owing to the fact that an appropriate experimental model for this disorder had not been developed. In the early 1970s such a model was developed in our laboratory (James *et al.*, 1974a, 1977).

Chronic communicating hydrocephalus can be induced in *Macaca fasicularis* monkeys by subarachnoid placement of a room-temperature mixture of Silastic (Dow Corning Chemical Corporation, Midland, MI). Although certain aspects of the technique have been previously reported, these will be summarized. The skin overlying the external occipital protuberance is shaved and aseptically prepared. Using this same midline landmark as a guide, a 17 gauge needle is inserted through a small incision and the needle advanced carefully to avoid the medulla. After appearance of clear CSF, a 19 gauge polyethylene catheter is introduced through the needle into the basal cistern. Free flow of CSF is used to ascertain proper catheter placement. Approximately 1 ml of a mixture of dimethyl polysiloxene, polysiloxene with filler and a catalyst (stannous octoate) is injected into the subarachnoid space. The animal is then left supine with its head in the dependent position for approximately 10 min to facilitate adequate flow cephalad and polymerization of the Silastic (Figure 13.1).

At sacrifice, all animals are anaesthetized with sodium pentothal, heparinized and transcardially perfused with Karnovsky's fixative (2.5% formaldehyde freshly depolymerized from paraformaldehyde and 1.0% glutaraldehyde in 0.1 M phosphate buffer, pH 7.2). The brains are removed and stored in the same fixative until samples are prepared for examination with transmission (TEM) and scanning electron microscopy (SEM).

The tissue is processed for SEM by modification of the OTOTO method. Samples are rinsed in 0.1 M phosphate buffer, pH 7.2, and post-fixed for 2 h in 1% OsO_4 in the same buffer. Six distilled water washes are followed successively by treatment with 1% TCH (30 min), 1% OsO_4 (2 h), 1% TCH (30 min) and 1% OsO_4 (2 h). Between TCH and OsO_4 treatments and after final incubation with OsO_4, samples are washed six times for 2.5 min each with distilled water and dehydrated through a graded ethanol series, critical-point dried from CO_2 and cemented to aluminium stubs with colloidal silver paste. Samples are then viewed without further coating in a Hitachi S-500 scanning electron microscope at 20 kV.

Specimens for TEM are dissected into blocks ($1 \times 1 \times 0.5$ mm) post-fixed in cacodylate-buffered 1% osmium tetroxide, rinsed in buffer, dehydrated in graded ethanols, stained *en bloc* with alcoholic uranyl acetate and embedded in Epon 812. Semithin (1.5 μm) sections are cut and stained with methylene blue and Azure II. The ultrathin sections are stained with

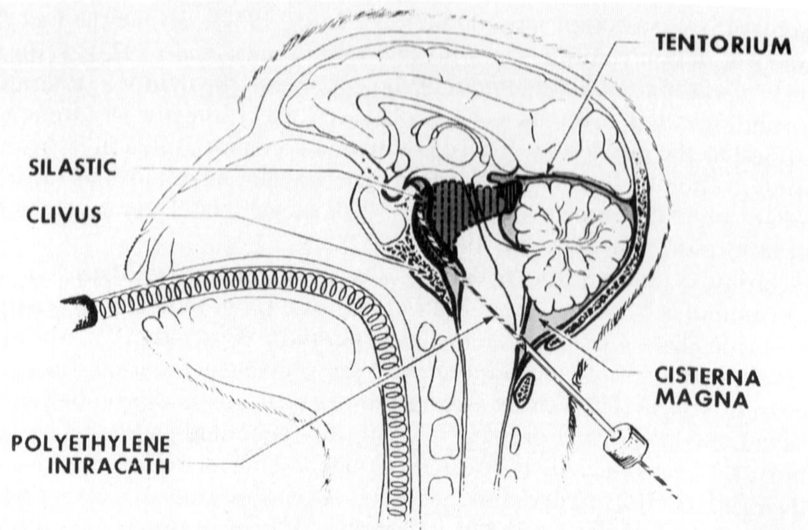

Figure 13.1 Drawing of animal model (rhesus monkey) for communicating hydrocephalus. Hatched area shows location of silicone rubber mixture at necropsy. Reproduced with permission from James *et al.* (1974a)

lead citrate and uranyl acetate and studied with a Philips EM 400 HMG TEM at 80 kV.

In many studies adjacent specimens for SEM are dissected into blocks which include the circumference of the lateral ventricle in the coronal plane, post-fixed in cacodylate-buffered 1% osmium tetroxide, dehydrated through graded ethanols and critical-point dried from liquid CO_2. Blocks are then mounted on stubs with silver or carbon paint, sputter coated with gold–palladium and examined with an ETEC Autoscan SEM at 20 kV. After SEM examination, selected blocks are rewetted with propylene oxide, infiltrated with Epon 812 and re-embedded for TEM.

EARLY ULTRASTRUCTURAL CHANGES

The lateral ventricular angle in the brain of the control monkey is too acute to permit examination of the angle vertex. In all of the experimental animals into which Silastic had been injected, the angle was dilated, permitting better assessment.

In animals studied from 12 h (acute) to 16 days (subacute) after Silastic injection, there were no obvious signs of tearing or rupturing of the ependymal surface throughout the 16 day period of study, but a number of very early cellular changes were observed.

Normally, the ependymal lining of the lateral ventricles is histologically made up of a cuboidal epithelium with bridging between the cells, thus maintaining the ependyma as a cellular layer separate from the ventricular cavity. Cilia and microvilli are found along the epithelial cells at the ventricular surface. The cross-sectional area of the lateral ventricles of the 16 day animal showed an approximately threefold increase in size when compared with the control with relatively little increase in surface area.

The ventricular surface appeared relatively smooth, but 'cobbling' was not seen in the acute animals (James *et al.*, 1980). No surface bulging could be seen on either the surface or the cross-sectional images. When sections were examined at the edges, no bulging of the ependymal nuclei was noted. The cells always presented a greater height than width.

These cross-sectional images of the ependymal cells were examined on the 5–6 day animals and were comparable with the control animal in height and width. Despite the focal denudation, some microvilli were clearly identified and appeared no different from the controls. There was a suggestion that they were heavier in areas between the ependymal cells. The ependymal cells, although stretched, appear to remain intact without being ruptured, and the cellular elements of the subependymal white matter are separated by a largely increased extracellular periventricular space.

The cilia of the ependymal cells at 12 h after Silastic injection appeared more densely packed than in the control (Figure 13.2). By 5 and 6 days,

(a)   (b)

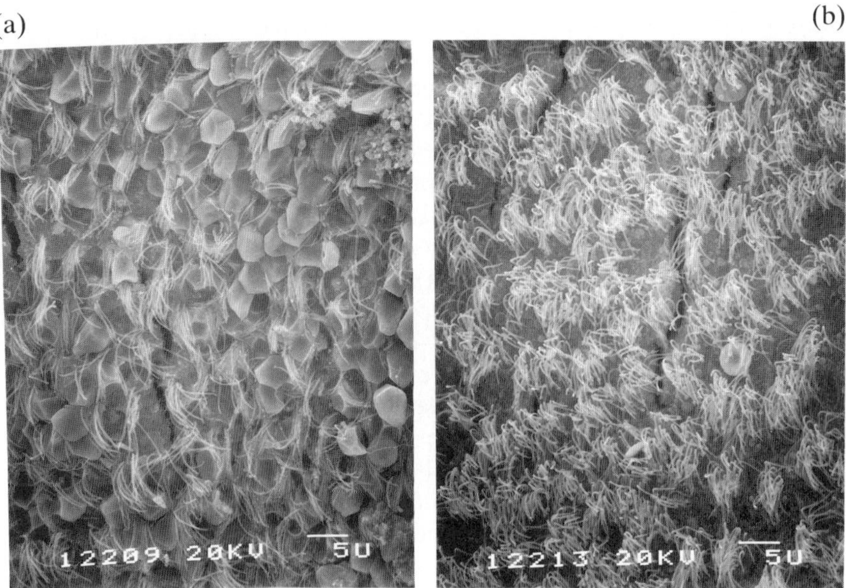

Figure 13.2 Superior surface of the lateral ventricle adjacent to the lateral angle. (a) Control primate. (b) Primate sacrificed at 12 h following injection of Silastic into the subarachnoid space. Note the relative density of the cilia as compared with the control. Bar = 5 μm

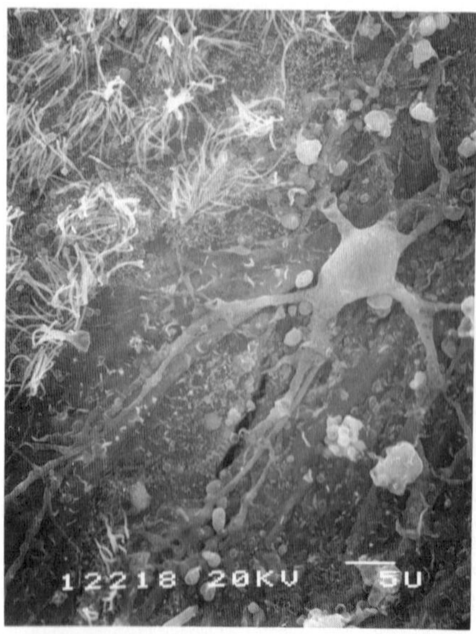

Figure 13.3 Superior surface of the lateral ventricle of a primate sacrificed at 5 days following Silastic injection. Note the extensive denuded areas and the appearance of SE cells. Bar = 5 μm

many of the cilia were flattened (Figure 13.3). At 16 days, this flattening was more pronounced and accompanied by clumping. Areas of cilia denudation, apparent at 12 h, increased with time (Figure 13.4). Coincident with the denuding was an increase in size and number of the supraependymal (SE) cells. The origin of these SE cells remains obscure. However, large fibrous astrocytes with a significant amount of glial filaments have previously been described, and it is believed that the supraependymal cells are of glial origin (James *et al.*, 1980). Although some of these cells may be obscured in the early stages by the dense cilia, numerous SE cells were apparent as early as 12 h. The SE cells, best seen in the denuded areas, were also present in the ciliated areas. These cells increased in size, number and complexity during the period studied and were similar to those observed in chronic preparations, which will be subsequently described in this chapter.

The microscopic changes in communicating hydrocephalus during the first 16 days include (1) an increase in density of cilia on the ependymal cells; (2) an increase in areas in which ependymal cells are denuded of cilia, and (3) an increase in number of supraependymal cells (Milhorat, 1970). Since denudation and the appearance of SE cells are known changes of chronic communicating hydrocephalus, these results indicate that anatomic and microscopic changes in communicating hydrocephalus occur as early as 12 h after the restriction of normal flow of CSF.

Ependymal cell flattening and focal denudation, as well as increase in

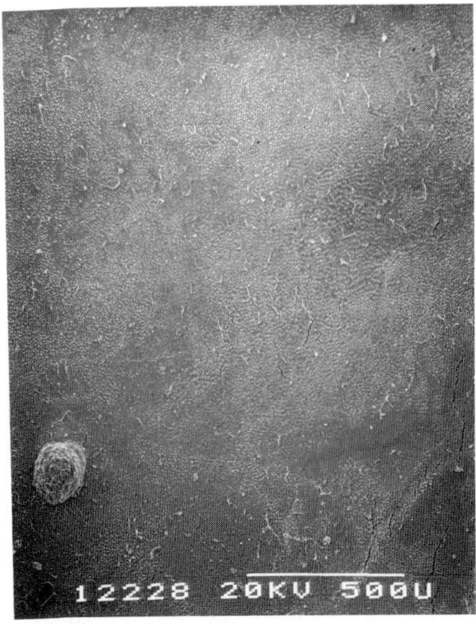

Figure 13.4   Denuded area of superior surface of the lateral ventricle from a primate sacrificed 16 days following Silastic injection. Bar = 500 μm

size and number of supraependymal cells, are ultrastructural changes that have been described by our group (Price *et al.*, 1976). These are shown in our experimental model 2 weeks after production of hydrocephalus. The cell populations of the CSF have not been studied, but one could theorize that with the increase in the extracellular field space in the periventricular region, any cell found in the CSF could be taken up in the periventricular white matter by transventricular absorption.

It has been postulated that as the intraventricular pressure slowly rises, the ependymal cells are progressively stretched, giving rise to alterations in their tensile properties. If the attenuation of these ependymal cells and the changes in the tensile properties are responsible for the clinical and pathological findings seen with so-called normal-pressure chronic communicating hydrocephalus, earlier confirmation of the ultrastructural changes should improve selection of those patients who would benefit from the surgical attempts at reversing their hydrocephalus.

LATE ULTRASTRUCTURAL CHANGES

We have also studied animals in the subacute and chronic stages of hydrocephalus from 6 to 1000 days and compared the acute, subacute and chronic preparations with normal.

The ventricular surfaces of the control animals contain many small microvilli and tufts of the larger cilia. These ciliary tufts are of sufficient density to appear as a 'lake bottom containing foliage'. The tufts move to and fro, effecting a 'current' in the CSF of the ventricle, later to be demonstrated by dynamic MRI. These cilia are more dense in the ventricular surface overlying the caudate nucleus than in the area over the corpus callosum. Microvilli are present over the ependymal surface but are concentrated at the cell borders, which clearly demonstrate the cell junctions. Large (6–10 μm) supraependymal cells are present in abundance with extensions branching over the ependymal surface. The ependymal cell layer is complete without obvious crevices or gaps and the cuboidal cells are regular in their appearance. The surfaces of the lateral ventricle in the dorsolateral angle betwen corpus callosum and caudate nucleus are completely covered with ciliated ependyma.

Despite the fact that there were no obvious signs of tearing or rupturing of the ependymal surface throughout the first 16 day period of study, a number of very early cellular changes were observed. The cross-sectional area of the lateral ventricles of the 16 day animal showed an approximately threefold increase in size when compared with the control, with relatively little increase in surface area. The ventricular surface appeared relatively smooth.

As noted, the cilia of the ependymal cells at 12 h after Silastic injection appeared more densely packed than in the controls. By 5 and 6 days, many of the cilia were flattened. At 16 days, this flattening was more pronounced and accompanied by clumping. Areas of denuded cilia, apparent at 12 hours, increased with time, and there was an increase in size and number of the supraependymal cells, the origin of which remains unknown. Large fibrous astrocytes with a significant amount of glial filaments are seen. Although some of these cells may be obscured in the early stages of development of hydrocephalus by the dense cilia, numerous SE cells were apparent as early as 12 h. The SE cells, best delineated in the denuded areas, were also present in the ciliated regions. These cells increased in size, number and complexity during the period studied.

Gross brain sections of the subacute animals 100 days after Silastic implantation demonstrate enlarged lateral ventricles with rounded and stretched angles. Normal ciliated ependyma are seen in the medial and lateral walls of the ventricles. However, the dorsolateral and dorsomedial ventricular angles demonstrate changes which are much more striking laterally. In these primates the ependymal cells progress from normal through a transitional zone of stretched and flattened profiles to frank denudation in the surface immediately adjacent to the angle. At 0.4–0.5 mm from the ventricular angle, the demarcation between the normal ciliated ependymal cells and the area of cell change or loss is well delineated. Numerous small (2.5–5 μm) rounded supraependymal (SE) cells are present in these areas of damage.

With TEM, these supraependymal cells are rounded or oval with a mottled electron-dense nucleus and a thin perinuclear rim of cytoplasm containing polyribosomes and an occasional mitochondrion. The SEM and TEM studies demonstrate the potential for direct communication between the ventricular lumen and the subependymal extracellular space through discontinuities up to 1 mm in the ventricular surface. These 'gaps' show that cell-to-cell apposition of the ependymal lining is not maintained in these regions. The network of subependymal astrocyte processes does not have the appearance of a significant barrier to passage of CSF from the ventricle into the brain parenchyma, explaining the observation at X-ray computed tomography of a 'halo' of diminished attenuation surrounding the ventricles. This circumstance may also be significant to the MRI CSF flow observations to be chronicled subsequently.

Primates that were studied 1000 days after Silastic placement demonstrated ventricles that were enlarged to a greater extent than in the primates studied after 100 days. In the area of the anterior horns, no ventricular septum was seen and the caudate nucleus remained as only a flattened rim along the ventrolateral border. The non-ciliated ventricular region in these chronic animals extended for a distance of approximately 10 mm from the dorsolateral and dorsomedial ventricular angles. The transition zone along the ependymal surface was again well demarcated as in the subacute animals. Even in the ciliated zones of the ventricle, however, the ependymal cells were noted to be flattened or stretched. The ependymal surface was more flattened and smooth than in the 100 day animals and in the region of severe pathology in the 1000 day animals was considerably different from that in the 100 day animals. Although unciliated, the cell covering of the ventricle was much more complete than that of the animals from the subacute group. Large (8–12 μm) flattened cell bodies with many fine, branching, radiating processes were present. These processes appeared in certain areas to spread out at the ends covering the underlying neural elements. Although the ependymal surface was better defined in the 1000 day animals, SEM and TEM studies showed multiple small openings or pathways into the brain parenchyma.

The small, round supraependymal (SE) cells that were abundantly demonstrated in great numbers in the animals studied at 100 days were not noted in the chronic hydrocephalic primates. Studies with transmission electron microscopy (TEM) showed that selected intercellular junctions, ependymal to ependymal as well as ependymal to astrocyte, had been re-established. In other areas hypertrophied processes of ependyma and astrocytes formed a loose meshwork similar to that seen in earlier animals. A potential pathway between the cerebral ventricular cerebrospinal fluid and the periventricular extracellular space appeared to remain in the chronic animals studied 1000 days after Silastic injection. This potential route of egress is further substantiated by our autoradiographic studies (James *et al.*, 1974b; Strecker *et al.*, 1974) and the appearance of the ventricular region on X-ray computed tomography.

We performed autoradiographic studies in the following manner (Bowsher, 1957). From 1 to 4 h prior to sacrifice, 100 μCi of ^{131}I-labelled albumin is injected into the subarachnoid space in the cisterna magna. After embedding the coronal sections in paraffin, 4 μm coronal slices through the lateral ventricles are obtained. The thin sections are placed on slides and coated with liquid emulsion (Kodak NTB2) which is uniformly in direct contact with the specimen. Slides are stored at 4 °C for a 24 h exposure time. They are then developed in D-19 (Eastman Kodak Corporation, Rochester, NY) and fixed. After washing, the specimen is dehydrated and covered. The autoradiographs are examined for silver grains in excess of background; their distribution and relation to the ependyma, cerebral veins and intracellular space are noted.

For the autoradiographic studies, coronal sections of the brain are made utilizing the following technique. The animal's brain is perfused with 500 ml of 10% formalin by injection into the carotid arteries at physiological pressures bilaterally. The brain and meninges are removed intact and placed in formalin for 24 h. Coronal sections of the brain are made and embedded in paraffin. Histological slides are then prepared and stained with haematoxylin and eosin.

In control animals (following slow intraventricular infusion of [^{131}I]-serum albumin), autoradiographs reveal little subependymal radioactivity and minimal periventricular radioactivity. Animals with communicating hydrocephalus demonstrate large amounts of radioactivity in the periventricular area after slow intraventricular infusion (Figure 13.5) (Strecker *et al.*, 1974). Semiquantitative and detailed autoradiographic data demonstrate facilitated diffusion (bulk flow) of the [^{131}I]-serum albumin and slowly decreasing concentration from the ependymal surface into the brain parenchyma in animals with chronic communicating hydro-cephalus. Radioactivity appears to be selectively oriented around the cerebral veins but no definite radioactivity is localized in the vein wall or noted intraluminally. Transfer studies of radioactivity from the CSF compartment to the intravascular space suggest that this may provide an alternative pathway for bulk flow of CSF (Bowsher, 1957; Strecker *et al.*, 1974).

These autoradiographic data suggest that in communicating hydrocepha-lus histological changes in the ventricular ependymal lining are accompa-nied by alteration in its permeability to large molecules. As albumin is a normal constituent of CSF, the transependymal migration of the labelled albumin suggests that the ventricle may participate as an alternative pathway of CSF absorption in chronic communicating hydrocephalus.

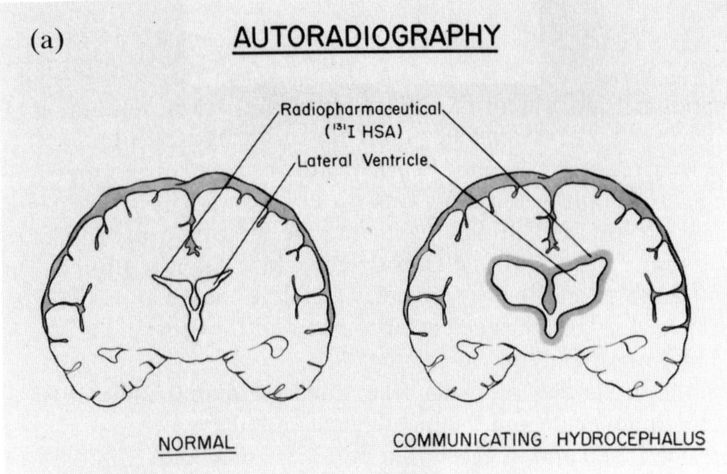

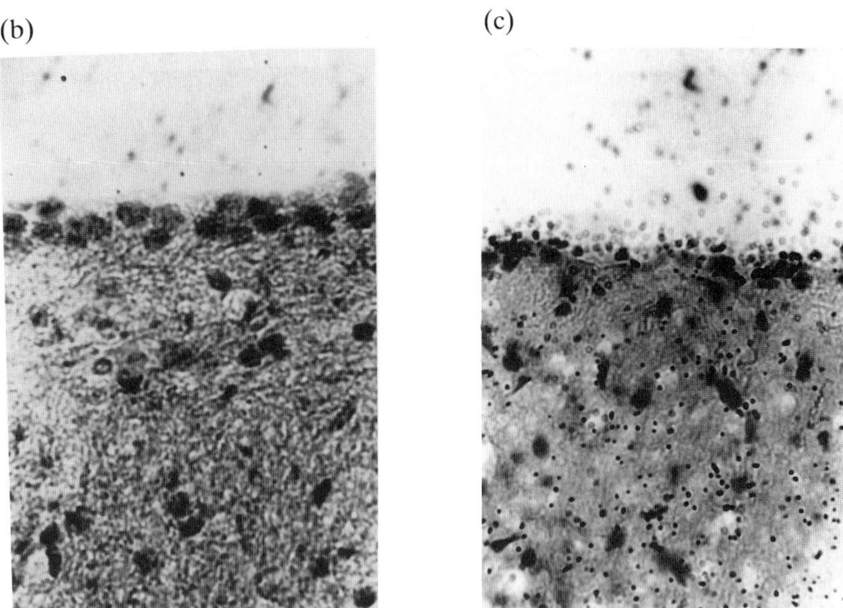

Figure 13.5 (a) Drawing of coronal section of autoradiographic findings in a normal animal and one with communicating hydrocephalus. (b) Autoradiograph in a normal dog after injection of [^{131}I]-serum albumin in the lateral ventricle ($\times$600 magnification, haematoxylin and eosin). The ependymal cells are of normal cuboidal shape. Few granules are present in the periventricular area. (c) Autoradiograph in a dog with communicating hydrocephalus. Ependymal cells lining the ventricle are flattened and in some areas appear denuded. Silver granules (reflected radioactivity) are present in the intercellular spaces in the periventricular areas in abundance. $\times$600 magnification, haematoxylin and eosin. Reproduced with permission from James *et al.* (1974b)

CLINICAL STUDIES

Transependymal movement of CSF may account for a number of clinical and diagnostic observations in communicating hydrocephalus. Lumbar subarachnoid injection of radiopharmaceuticals at cisternography reveals ventricular entry and retention. This observation suggests a 'reversal' of the normal currents of CSF flow, allowing the radiopharmaceutical to pass from the basal cisterns into the lateral ventricles. The retention or stasis of the labelled substances in the ventricles could be caused by diminished net outward flow due to enlargement accompanied by ventricular absorption initiated by transependymal migration.

Permeability changes allowing free transependymal movement of the large albumin molecule and the histological and ultrastructural evidence of periventricular oedema suggest that the CSF may move into the brain parenchyma by a process more dynamic than that of simple diffusion. The decreased attenuation of the X-ray beam at computed tomography allows us to make this observation *in vivo* in a non-destructive manner. As we present the MRI observations of CSF physiology, the importance of the increased ECS in the periventricular region will also become evident. The issue of periventricular compliance may well be addressed by the findings in our experimental model.

Computed Tomography (CT)

Within a few years of its introduction as a clinical capability, X-ray computed tomography (CT) had supplanted pneumoencephalography (PEG) as a means of studying intracranial pathological processes. Initially, the significant advantage of CT over PEG was the ability to rapidly and non-invasively assess ventricular size and displacement. As CT technology matured, it became possible to image finer anatomic details and resolve smaller differences in attenuation of the X-ray beam. This permitted, for example, the detection of a low-density 'halo' around the ventricles, which was believed to represent transependymal migration of CSF in the instance of both intraventricular and extraventricular CSF obstruction. With the introduction of water-soluble contrast agents which could safely be used intrathecally, CT was able to provide some degree of physiological information about CSF flow dynamics.

The two main categories of communicating hydrocephalus, communicating hydrocephalus *ex vacuo* and communicating hydrocephalus secondary to extraventricular obstruction, are regarded as having different appearances on CT scanning.

Hydrocephalus *ex vacuo* is a generalized atrophic process resulting in a symmetric increase in size of the ventricles, cisterns and sulci over the convexity and is readily apparent on either CT or MR, rarely requiring additional imaging evaluation (Figure 13.6). In these cases of atrophy radionuclide cisternography studies show delayed movement through an enlarged subarachnoid space (Partain and Staab, 1979).

Communicating hydrocephalus due to extraventricular obstruction usually follows subarachnoid haemorrhage, trauma or an inflammatory process. It results from an imbalance between CSF production and absorption, usually from impaired absorption due to inflammatory changes and repair at the base of the brain, the tentorial hiatus, or over the cerebral convexities (Quencer *et al.*, 1989). In these patients, CT will show ventricular enlargement which is more severe than the amount of basilar cistern enlargement or prominence of the cortical sulci (Figure 13.7).

Efforts to further the utility of CT in the assessment of cerebral atrophic and obstructive disorders have met with little success. Quantitative analysis of the compliance of the ventricles has been attempted by comparing the X-ray attenuation numbers of the ventricles of normals with those of patients with communicating hydrocephalus. This analysis did not accurately separate the two groups (Di Chiro, G., personal communication).

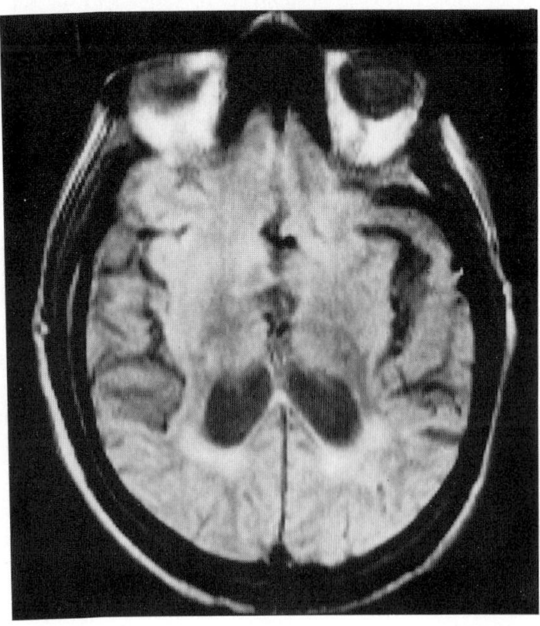

Figure 13.6 Transverse MRI image of midbrain cortical loss (spin echo 2000/30). The ventricles are enlarged, as are the cortical sulci. Reproduced with permission from Partain, C. L., Sandler, M. P., Price, A. C. and James, A. E., Jr. (1989). Cerbrospinal fluid imaging. In Sandler, M. P., Patton, J. A., Shaff, M. I., Powers, T. A. and Partain, C. L. (Eds), *Correlative Imaging: Nuclear Medicine, Magnetic Resonance, Computed Tomography, and Ultrasound*. Williams and Wilkins, Baltimore, pp. 175–191

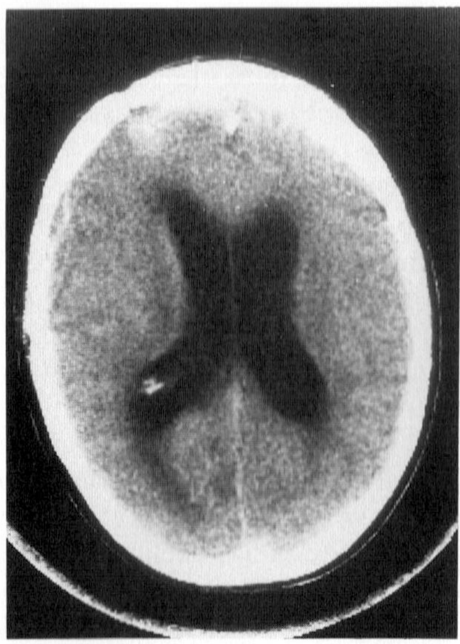

Figure 13.7   Hydrocephalus in a patient several months following a subarachnoid haemorrhage. Note lack of convexity sulci definition in addition to the ventricular enlargement. There are several areas of infarcts, which are complications of the subarachnoid haemorrhage. Reproduced with permission from Partain, C. L., Sandler, M. P., Price, A. C. and James, A. E., Jr. (1989) Cerebrospinal fluid imaging. In Sandler, M. P., Patton, J. A., Shaff, M. I., Powers, T. A. and Partain, C. L. (Eds), *Correlative Imaging: Nuclear Medicine, Magnetic Resonance, Computed Tomography, and Ultrasound*. Williams and Wilkins, Baltimore, pp. 175–191

Metrizamide-enhanced dynamic CT has been used for CSF flow measurement (Partain *et al.*, 1978; Enzmann *et al.*, 1979) and has been described as having potential in improving diagnostic criteria for patients with communicating hydrocephalus. However, neither metrizamide-enhanced CT scans nor radionuclide CSF flow studies (cisternography) have proven to be consistent in predicting the clinical outcome of ventricular shunting (Partain *et al.*, 1978; Enzmann *et al.*, 1979).

At present, CT remains an excellent means for assessing ventricular size and displacement, for assessing presence of CSF in the periventricular white matter (James *et al.*, 1974b) and for detecting additional morphological information (such as associated sulci enlargement). It is not a method either for reliably diagnosing communicating hydrocephalus or for determining which patients will benefit from diversionary shunting of CSF. However, information provided by new methods in magnetic resonance imaging, combined with the traditional modalities, may prove to be more successful.

Magnetic Resonance Imaging (MRI)

Magnetic resonance imaging (MRI) has several advantages over X-ray computed tomography. Dilation of the ventricular system is easily diagnosed by MRI without ionizing radiation or any other significant known biological hazard. In CT, images represent attenuation measurements from slabs of tissue typically 4–8 mm thick. Since more than one type of tissue may be present in a given plane, the resulting X-ray attenuation number is an average of those residing in the individual tissues. Because of this limitation, the attenuation from slabs which contain both CSF and brain can be confused with periventricular low X-ray attenuation (Randall *et al.*, 1983; Partain *et al.*, 1988a). For this reason and because MRI is more sensitive than CT to small amounts of extraventricular water, MRI is preferred to CT for visualization of periventricular oedema. This capability allows the same differentiation of hydrocephalus due to obstruction from that due to atrophy that CT does, only with more sensitivity (Figure 13.8). The MRI criteria for communicating hydrocephalus include a thin uniform periventricular rim of abnormally increased signal intensity on T2-weighted images and the same morphological features described earlier by CT (see above).

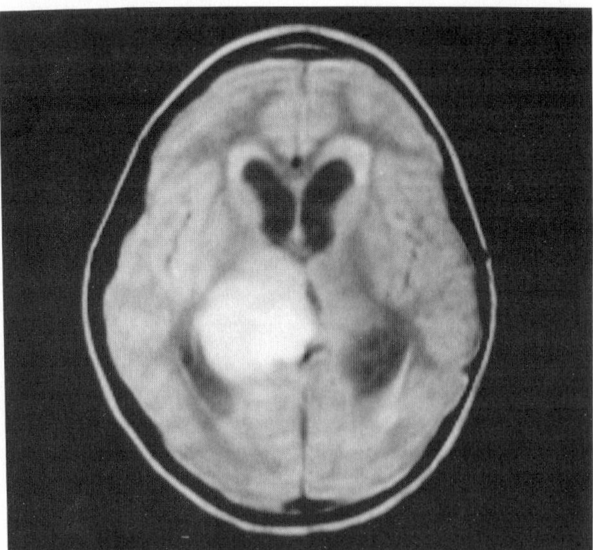

Figure 13.8 Acute obstructive hydrocephalus (due to a thalamic tumour obstructing the cerebral aqueduct) with periventricular oedema seen as a high signal intensity on spin echo (2000/60) technique (transverse slice). Reproduced with permission from Partain, C. L., Sandler, M. P., Price, A. C. and James, A. E., Jr. (1989). Cerebrospinal fluid imaging. In Sandler, M. P., Patton, J. A., Shaff, M. I., Powers, T. A. and Partain, C. L. (Eds), *Correlative Imaging: Nuclear Medicine, Magnetic Resonance, Computed Tomography, and Ultrasound.* Williams and Wilkins, Baltimore, pp. 175–191

Qualitative Evaluation of Flow: Signal Voids Due to Flow

Information on CSF dynamics can be inferred from spin-echo studies or can be visualized and quantified directly with dynamic MRI techniques. The first use of MRI in evaluating CSF dynamics was in the detection of the presence of signal voids due to flow in spin-echo images. Signal intensity in MR images is affected both by random microscopic motion (diffusion) and macroscopic flow. Random microscopic motion results in irreversible dephasing of the transverse magnetization, causing a signal loss (Wesley *et al.*, 1984). Slow laminar flow results in signal loss on the first echo of a multiecho sequence but an increase in signal on the second and subsequent even-numbered echoes due to rephasing of the magnetization (Waluch and Bradley, 1984).

Sherman and Citrin (1986) observed the flow void phenomenon in the aqueduct of Sylvius on T2-weighted images when compared with the signal in the lateral ventricles. The flow void phenomenon was seen in some patients on T1-weighted images as well. Bradley *et al.* (1986) found flow voids in the cerebral aqueduct of normal volunteers, but noted that an area of signal dropout in the aqueduct was most pronounced in patients with chronic communicating hydrocephalus, less evident in patients with acute communicating hydrocephalus and least evident in patients with atrophy (Figure 13.9). Since ventricular compliance is believed to be essentially normal in atrophy, mildly decreased in acute communicating hydrocephalus and severely decreased in normal-pressure hydrocephalus (NPH), the degree of signal loss may reflect the velocity of the pulsatile CSF motion, which is dependent on the relative ventricular compliance and surface area (Bradley *et al.*, 1986).

The use of the aqueductal flow void sign to diagnose NPH has been extended (Bradley *et al.*, 1989) to be used as a predictor for the response to ventriculoperitoneal shunting, by retrospectively studying 20 patients. Patients were graded, using a double-blind method, by a neurologist or neurosurgeon as 'excellent', 'good', 'fair' or 'poor' responders to shunting. The degree of CSF flow void in the aqueduct and fourth ventricle on T2-weighted images and the degree of deep white matter infarction was compared with the clinical results following shunting. The patients with a marked CSF flow void had a significantly better outcome than those with normal or reduced flow voids.

Phase Reconstructions in Spin-echo Imaging

Another method used to evaluate CSF dynamics has been phase reconstruction(s) of velocity-compensated spin-echo images. Phase reconstructions can be used for both qualitative and quantitative analysis, although

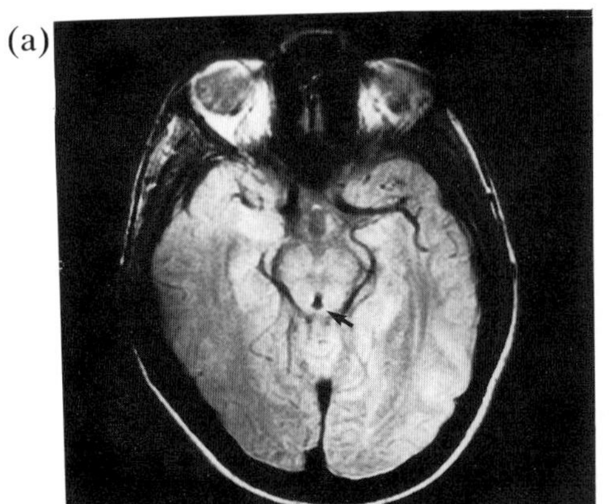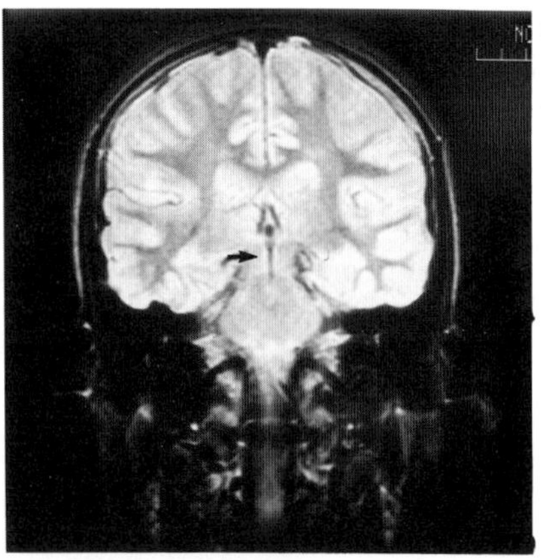

Figure 13.9   The aqueductal flow void as imaged in a normal patient. (a) This single axial MR image shows a flow void (arrow) in the aqueduct of Sylvius. (b) In this coronal MR image, the entire length of the aqueduct (arrow) is demonstrated as a flow void

qualitative techniques are currently utilized more frequently because of the careful calibration of sequences required for accurate quantitative results. The phase shift in an image will be linearly related to the gyromagnetic ratio, the strength of the flow-sensitizing gradient used, the duration of these gradients and the velocity of the protons (Moran, 1982):

$$\Delta\phi = \gamma G d v D$$

where $\Delta\phi$ = phase shift; γ = gyromagnetic ratio; G = gradient strength; d = duration of the gradient; v = velocity of the proteins; and D = time interval between flow-sensitizing gradients. In a system which has been calibrated using flow phantoms, the direction and velocity of flowing material can therefore be determined from the signal intensity. Stationary material will appear grey and isointense on the images, protons flowing out of plane will be white and those flowing into the plane will be black (Figure 13.10). If the gradient is applied in the opposite direction, the intensity scale will be reversed.

Several investigators have used phase reconstructions qualitatively to assess flow patterns of CSF without quantifying the magnitude of CSF flow. Using ECG gating, multiple spin-echo images through the cardiac cycle can be obtained. Levy *et al.* (1988b) studied both normal volunteers and patients. They found flow reversal in the posterior versus the anterior portions of the spinal canal, in the cerebellomedullary cistern relative to cervical flow, near the walls of the fourth ventricle and in the lumbar area relative to the cervical region. Flow into known communicating cysts or cavities of the CSF space was seen. Nadel *et al.* (1989) found phase images superior to T2-weighted magnitude spin-echo images for assessing the

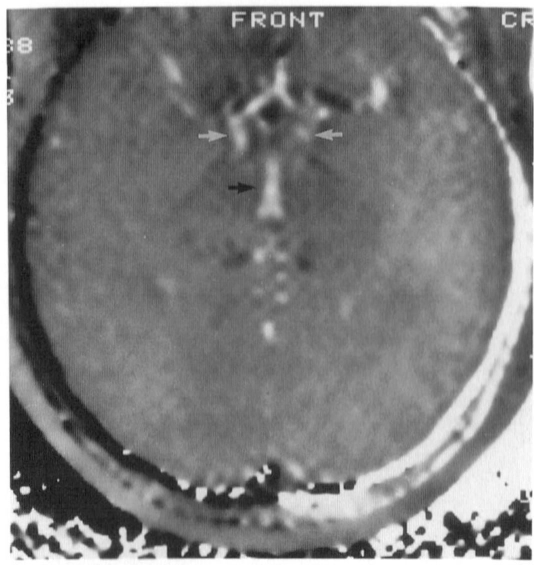

Figure 13.10   Phase-reconstructed image of CSF flow. In this axial MR section, motion is encoded such that flowing CSF and blood are white or black, and areas with no motion are grey. Pulsatile CSF motion is best seen in the third ventricle (black arrow) and the paired foramina of Monro (white arrows)

presence or absence of flowing blood in patients with patent and thrombosed superior sagittal sinuses, AVMs and aneurysms, owing to the variable appearance of blood in a clot depending on its age. In addition, they noted that phase imaging of CSF flow was also more sensitive than long T2-weighted images.

Ridgway *et al.* (1987) quantified flow using phase reconstructed spin-echo images. They measured flow velocities and volume flow rates in volunteers at the level of C2 using images made at several points in the cardiac cycle. Ciraolo and Mascalchi (1988) have also used spin-echo phase images to quantitatively assess CSF flow. Using images acquired in diastole and systole of a thin sagittal plane through the aqueduct of Sylvius, they evaluated normal volunteers and neurological patients with and without communicating hydrocephalus. In normals and in patients without communicating hydrocephalus, they found retrograde flow in diastole from the fourth to the third ventricle and antegrade flow in systole. After calibration with a phantom, CSF velocity in the aqueduct was approximately 5 mm/s. In communicating hydrocephalus, there was permanent retrograde flow even in systole.

Martin *et al.* (1989) studied the use of flow measurements with spin-echo phase images for the evaluation of shunt patency in patients with hydrocephalus. A special radiofrequency (RF) coil was designed in order to obtain accurate quantitative results in the small field of view in the shunt (1.2 mm). The technique, which produces three-dimensional maps of velocity similar to that shown in Figure 13.11, was validated in a phantom for various flow rates. Flow rates as low as 2 ml/h were measured accurately with this technique and the technique was proven to be clinically feasible (Martin *et al.*, 1989).

The clinical utility of phase-reconstructed spin-echo images at present consists first of its qualitative ability to show directionality of flow. The second area of utility is the quantitative measurement of flow and the analysis of cyclical changes of these instantaneous flow rates in phase with the cardiac cycle. Unfortunately, these techniques have a long acquisition time; given the elderly patient population usually having such examinations, there is often serious degradation of image quality due to patient motion. Newer, faster MR sequences hold promise for more reliably obtaining clinically useful information from the typical patient.

Development and Utility of Cine Techniques

Since CSF pulsation is related to the cardiac cycle (du Boulay, 1966), it is desirable to acquire images at multiple time points within the cycle for a complete analysis of CSF dynamics. However, assessing multiple points in the cardiac cycle at the same physical slice location with spin-echo sequences is time-consuming. In the last several years, faster sequences have been developed which allow acquisition at multiple points in the cardiac cycle in under 5 min (Edelman *et al.*, 1986; Jolesz *et al.*, 1987). These sequences, including gradient echo and steady-state free precession (SSFP) sequences, can obtain multiple images at the same location at

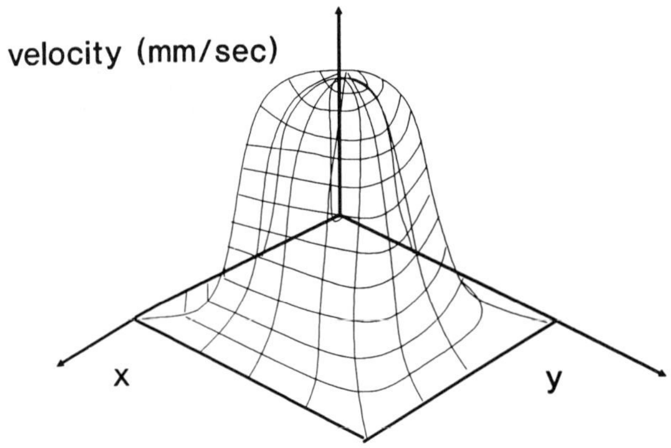

Figure 13.11 Three-dimensional velocity map of shunt flow

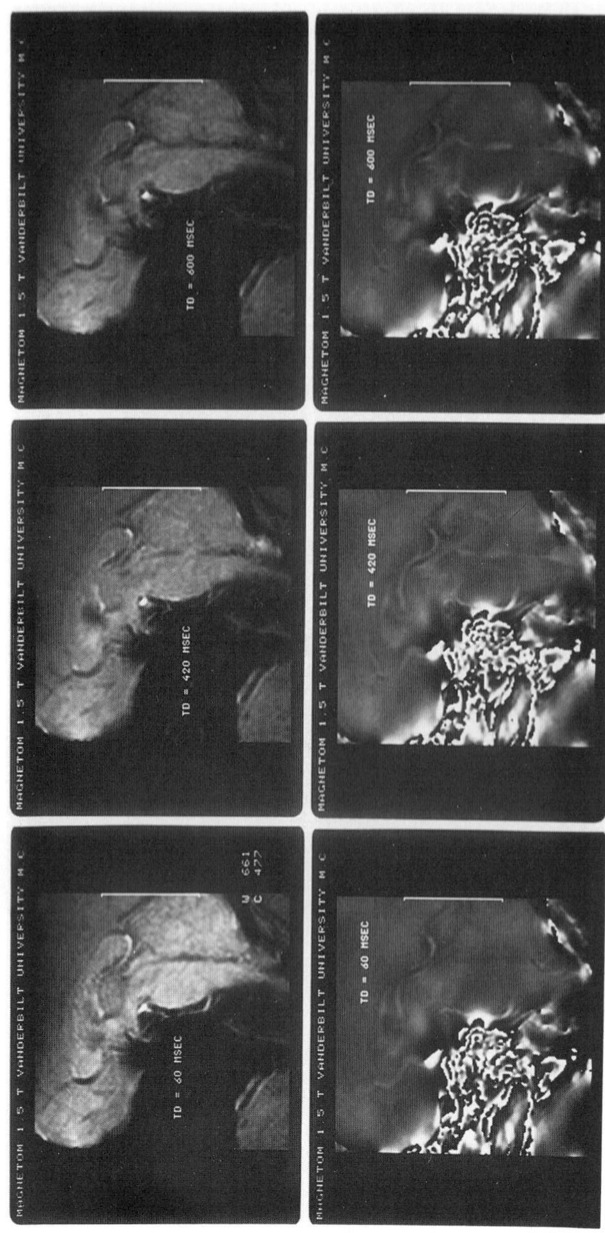

Figure 13.12 Images of cerebral aqueduct produced with a gradient echo cine sequence. Magnitude images at various points in the cardiac cycle show low signal in systole (A = 60 ms from R-wave), higher signal near end-systole (B = 420 ms) and moderate signal in diastole (C = 600 ms). The subject's R–R interval was approximately 1 s. The corresponding phase-reconstructed images show flow reversal between 60 and 420 ms (D and E) and slightly faster flow (higher signal) in diastole (F—600 ms). (Images run left to right in each row: A = top left; D = bottom left)

(a)

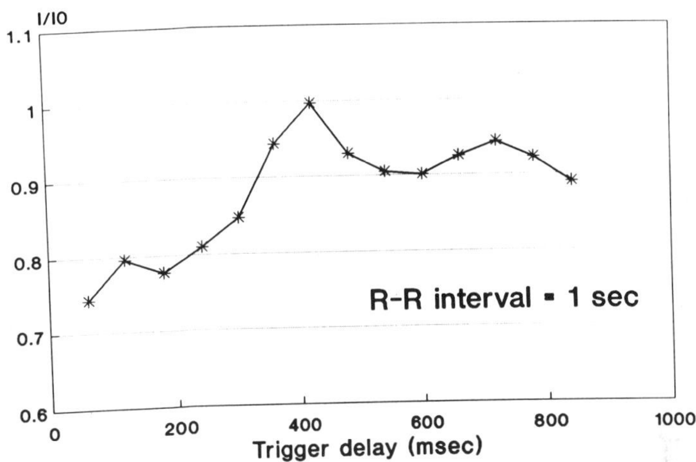

(b)

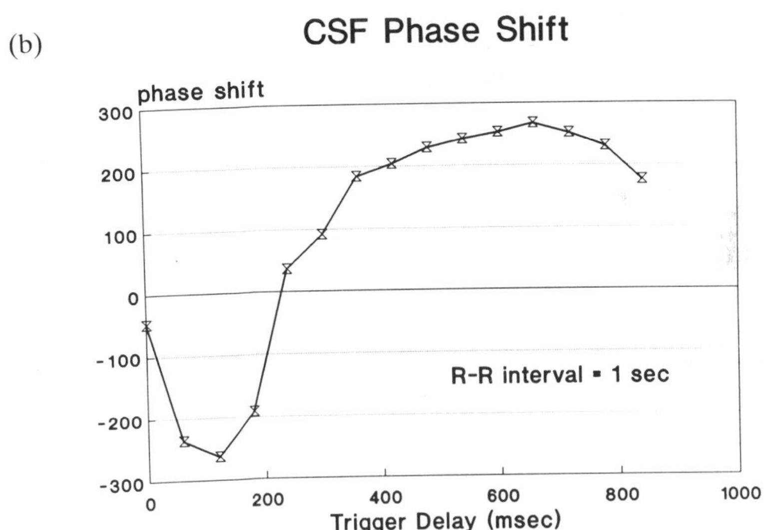

Figure 13.13 (a) The ratio between signal intensity in the aqueduct and signal intensity in the cerebral pons is plotted as a function of time after the R-wave for the images shown in Figure 13.12. The intensity is related to the velocity of the CSF flow, but the direction of the flow cannot be determined. (b) Both magnitude and direction of CSF flow can be determined from the signal intensity of the phase-reconstructed images. The sign of the signal changes from negative to positive in systole (200 ms), indicating a reversal in the direction of flow from cranio–caudal to caudal–cranial

different phases of the cardiac cycle, all in one acquisition. Since the collected images can be stored and then viewed rapidly in sequence to create a movie loop, they are often collectively referred to as cine sequences. These cine sequences can be designed and calibrated to measure velocity as in spin-echo imaging (Figures 13.12, 13.13). By using the signal intensity of the magnitude images and the intensity of the phase reconstructions, investigators have attempted to characterize CSF flow quantitatively and qualitatively.

Much of the work to date has been to qualitatively characterize normal and abnormal CSF flow patterns using cine sequences (Harada *et al.*, 1988; Ida *et al.*, 1988; Kawahara *et al.*, 1988; Quencer *et al.*, 1989). Kawahara *et al.* (1988) studied CSF flow in patients with hydrocephalus, patients with atrophy and normal volunteers with cine MRI, and qualitatively analysed signal intensity and direction of flow. In their cine sequence high flow was imaged as an increase in signal. They observed increased flow in the third and fourth ventricles, especially in those patients with communicating hydrocephalus. The signal intensity in the aqueduct was higher (denoting flow) than the midbrain in controls, and patients with NPH and with atrophy, while the signal was lower (denoting less flow) than the midbrain in patients with obstructive hydrocephalus.

Increasingly more attention has been given to quantitative measurements in the last few years as the techniques for designing and calibrating cine sequences have evolved. Ciraolo *et al.* (1988) have compared their results from multiple ECG-gated spin-echo images (Nadel *et al.*, 1989) with those obtained by analysing signal–time curves from the magnitude images obtained with a cine sequence. They normalized the signal of the flowing CSF to that of stationary CSF and found good correlation with their previous results. Enzmann *et al.* used phase images to establish the normal velocity–time curves for flow in the cervical spine subarachnoid space for comparison with patients with syringomyelia (Enzmann *et al.*, 1989; Van den Hout *et al.*, 1989). In these studies velocity, calculated from the phase images, and signal intensity were plotted as a function of time. Good correlation was found between the two methods of determining velocity profiles. Rubin found that the presence of signal loss on T2-weighted spin-echo images was not direction sensitive and did not correlate with either the magnitude or direction of CSF motion measured with a cine sequence as they evaluated patients with syringomyelia (Rubin and Enzmann, 1988). Mascalchi *et al.* (1989) evaluated aqueductal CSF flow in demented patients who also had hydrocephalus, other demented patients and normal volunteers. The signal in the aqueduct was normalized to stationary CSF (I/I_0). By plotting this ratio as a function of time, the periodic velocity profiles were mapped. The mean maximum I/I_0 was significantly greater in patients with communicating hydrocephalus than that found for the other groups. No significant difference was found between the mean minimum values for the three groups. Mascalchi *et al.* conclude that the evaluation of aqueductal CSF flow as determined by cine MRI may assist in differentiating dementia conditions implying hydrocephalus on pathophysiological grounds.

Van den Hout *et al.* (1989) used signal intensity curves to evaluate normal volunteers and patients. They used signal intensity–time curves to determine the point in time after the R-wave at which reversal of flow in the cerebral aqueduct occurs. In the patients studied, those with complete aqueductal obstruction showed little signal change throughout the cardiac cycle. All others, including those with communicating hydrocephalus, exhibited the same shape profiles as those of the normal subjects. However, the patients with communicating hydrocephalus had significantly higher signal intensity than the normal subjects. Signal intensity in these sequences is related to velocity of CSF through the aqueduct, but the exact relationship is complex and not fully understood. At the present time, the only manner in which one can accurately map velocities is with the use of calibrated phase images. No phase images were used in this study to allow direct calculation of velocity. They conclude that investigation into factors such as aqueduct diameter, CSF volume and brain compliance, combined with direct flow measurements, will be necessary before the meaning of signal intensity curves is completely understood.

Feinberg and co-workers used a variation on traditional cine techniques to create velocity density images from a narrow slab of tissue with a Fourier transform technique (Feinberg *et al.*, 1985; Feinberg and Mark, 1987). In these images velocity is plotted in the vertical direction and spatial position is plotted in the horizontal direction. The method was found to be sensitive to changes in velocity as small as 0.4 mm/s. From application of the technique to both brain and CSF motion, the authors proposed that a vascular-driven movement of the entire brain may be pumping the CSF circulation (Feinberg and Mark, 1987).

Others have used Fourier analysis to obtain further information from the velocity–time curves calculated from phase images (Thomsen *et al.*, 1989). The velocity–time function was Fourier transformed, and the zeroth, first, second and third harmonics were determined. The zeroth harmonic (offset) was used to calculate flow production in phantoms and in a normal volunteer. In a phantom with no net flow, the offset was zero, indicating equal inflow and outflow. In the normal volunteer, a net production of 12 mm^3 per cardiac contraction (1.1 l/day) was found. This method shows promise for the quantification of CSF production.

Cine techniques, in summary, have confirmed known principles of CSF flow, such as pulsatile motion, dependence on the cardiac cycle, and even production rates of CSF. In clinical use, these techniques will likely permit both the more reliable diagnosis of communicating hydrocephalus, and a more accurate prediction of which patients will respond favourably to ventricular shunting.

Bolus-tracking Techniques

Recently several new approaches to evaluating CSF dynamics with MRI have been implemented. These methods, referred to as bolus-tracking methods, have evolved from traditional MR imaging techniques (Shimizu *et al.*, 1986; Levy *et al.*, 1988c, 1990; Edelman *et al.*, 1989; Kahn and Mueller, 1989; Kraft *et al.*, 1990). These methods are technically demanding and, therefore, have not yet been perfected. However, they hold promise for more accurately quantifying CSF flow and forces in the surrounding tissues than current methods. It is for this reason that they are described here.

In bolus-tracking techniques slabs of fluid and/or tissue are selected by varying their magnetization prior to imaging. The simplest method is a projection technique in which a selected plane orthogonal to the vessel or aqueduct is used to create one-dimensional images where the intensity is related to the line integral along the plane (Shimizu *et al.*, 1986). This intensity will vary in time as the fluid within the plane pulsates, and can be related to velocity.

In the presaturation technique (Edelman *et al.*, 1989) a slab perpendicular to the plane of the vessel is selected and tagged. The movement of this tag as a function of time is related to flow velocity (Figure 13.14). This method has an advantage over the projection method in that it preserves visualization of the CSF anatomy. However, it is inconvenient to tag multiple sections in an image using this technique.

In the spatial modulation of magnetization technique (SPAMM) (Levy *et al.*, 1988a; Axel and Dougherty, 1989) multiple slabs in the image plane are selected and tagged prior to imaging. The appearance of the tags in the resulting images allows for the quantification of flow (Figure 13.15). Levy *et al.* (1990) used this method to create a series of stripes across the cervical spinal cord. Serial images obtained at different phases of the cardiac cycle demonstrate deflections of the stripes and correlated with measurements based on spin-echo phase images and presaturation tagging (Levy *et al.*, 1988c). Additionally, internal deformation within the cord was noted, demonstrating the effect of differential and shear forces within the cord. The SPAMM technique was found to be faster than the presaturation technique in terms of stripe placement. It also allowed placement of more stripes over the field of view, which resulted in improved temporal and spatial resolution over presaturation techniques. Kraft *et al.* (1990) have reviewed all three methods, and found that the choice of technique should

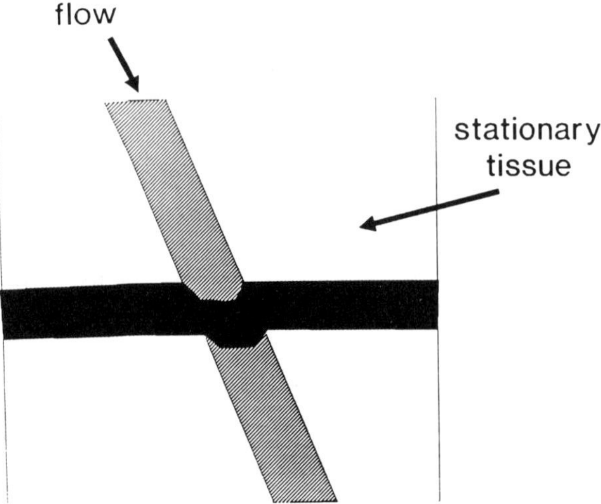

Figure 13.14   Presaturation technique. A slab of tissue is saturated prior to imaging and produces no signal. As the CSF moves through the slab, its shape is distorted. The extent of this distortion can be mapped as a function of time and related to flow velocity and direction

depend on the range of velocities expected. They conclude that the SPAMM technique may prove to be the most useful of the three for evaluating CSF flow.

SUMMARY

Communicating hydrocephalus and related disorders constitute an ever-increasing series of clinically significant problems. The basic experimental data provided by Davson and others have established fundamental principles to understand the underlying functional abnormalities that produce the pathophysiological expressions of the important central nervous system diseases associated with hydrocephalus and to assist in differentiating these treatable processes from other forms of dementia.

The authors have attempted to relate a body of their research in hydrocephalus and that of others to the various imaging modalities present in our clinical armamentarium. Particularly, we have emphasized dynamic magnetic resonance imaging, as we believe it holds great future promise in the evaluation of patients with dementia and, more specifically, communicating hydrocephalus associated with dementia. The legacy of our colleagues in the basic sciences is clearly evident when we analyse these unique images provided by the biomedical engineering advance of magnetic resonance.

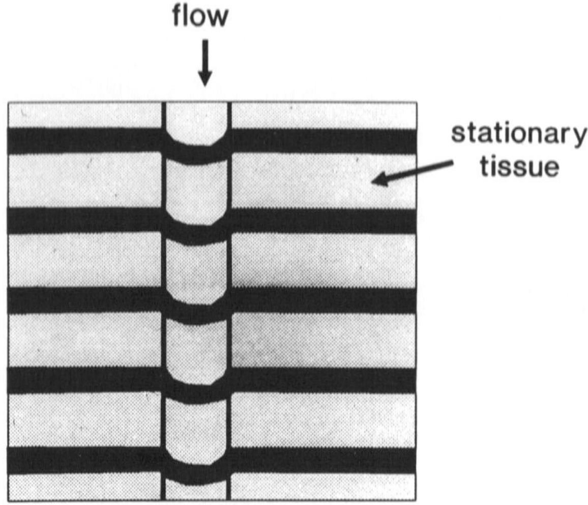

Figure 13.15 Spatial modulation of magnetization technique (SPAMM). Multiple slabs of tissue are saturated prior to imaging. The advantage of the SPAMM technique over the presaturation technique is that the placement of multiple tagged planes across the entire range allows a more complete assessment of CSF flow velocity and direction

ACKNOWLEDGEMENTS

We are indebted to Hugh Davson for his advice and counsel and for the opportunity for the senior author to conduct CSF research in his laboratory at University College under the funding of the Royal Society of Medicine Foundation. Gordon McComb, Jillian Christensen, Gary Novak, Bill Flor, Giovanni Di Chiro, Don Price, Mel Epstein, John Harbert, Guy McKhann, Barry Burns, Joe Diggs and others have collaborated with us on individual research projects. Funding was provided by the Picker Foundation, the National Radiation Council, the National Academy of Science, the National Institutes of Health Neuroradiological Training Grant, the National Foundation, the Royal Society of Medicine Foundation, Biomedical Research Support Grant of Vanderbilt University, the Vanderbilt Center for Medical Imaging Research and the Siemens Corporation.

REFERENCES

Axel, L. and Dougherty, L. (1989). MR imaging of motion with spatial modulation of magnetization. *Radiology*, **171**, 841–845

Bowsher, D. (1957). Pathways of absorption of protein from the cerebrospinal fluid: an autoradiographic study in the cat. *Anat. Rec.*, **128**, 23–29

Bradley, W. G., Kortman, K. E. and Burgoyne, B. (1986). Flowing cerebrospinal fluid in normal and hydrocephalic states: appearance on MR images. *Radiology*, **159**, 611–616

Bradley, W. G., Whittemore, A. R., Kortman, K. E., Caton, W. L. and Garner, J. T. (1989). Significance of the aqueductal cerebrospinal fluid flow void and deep white matter infarction on the outcome in patients with shunts for normal pressure hydrocephalus. *Radiology*, **173**(P), 115

Ciraolo, L. and Mascalchi, M. (1988). Cerebrospinal fluid flow quantification in the aqueduct of Sylvius with cardiac-gated MR phase imaging. *Radiology*, **169**(P), 211

Ciraolo, L., Mascalchi, M. and Bucciolini, M. (1988). Comparative (spin-echo/field-echo) MR study of cerebrospinal fluid dynamics in the Sylvian aqueduct. *Radiology*, **169**(P), 343

Davson, H., Kleeman, C. R. and Levin, E. (1962). Quantitative studies of the passage of different substances out of the CSF. *J. Physiol. (London)*, **161**, 126–142

Davson, H., Welch, K. and Segal, M. B. (1987). *The Physiology and Pathophysiology of the Cerebrospinal Fluid*. Churchill Livingstone, Edinburgh

DeLand, F. H., James, A. E., Ladd, D. J. and Koningsmark, B. W. (1972). Normal pressure hydrocephalus: a histological study. *Am. J. Clin. Pathol.*, **58**, 58–63

Diggs, J., Price, A. C., Burt, A. M., Flor, W. J., McKanna, J. A., Novak, G. R. and James, A. E., Jr. (1986). Early changes in experimental hydrocephalus. *Invest. Radiol.*, **21**, 118–121

du Boulay, G. H. (1966). Pulsatile movements in the CSF pathways. *Br. J. Radiol.*, **39**, 255–262

Edelman, R. R., Mattle, H. P., Kleefield, J. and Silver, M. S. (1989). Quantification of blood flow with dynamic MR imaging and presaturation bolus tracking. *Radiology*, **171**, 551–556

Edelman, R. R., Wedeen, V. J., Davis, K. R., *et al.* (1986). Multiphasic MR imaging: a new method for direct imaging of pulsatile CSF flow. *Radiology*, **161**, 779–783

Enzmann, D. R., Norman, D., Price, D. C., *et al.* (1979). Metrizamide and radionuclide cisternography in communicating hydrocephalus. *Radiology*, **130**, 681–686

Enzmann, D., Rubin, J. and Pelc, N. (1989). Cine phase contrast maps of cervical cerebrospinal fluid motion. *Radiology*, **173**(P), 157

Feinberg, D. A., Crooks, L. E., Sheldon, P., Hoenninger, J., Watts, J. C. and Arakawa, M. (1985). Magnetic resonance imaging the velocity vector components of fluid flow. *J. Magn. Reson. Med.*, **2**, 555–556

Feinberg, D. A. and Mark, A. S. (1987). Human brain motion and cerebrospinal fluid circulation demonstrated with MR velocity imaging. *Radiology*, **163**, 793–799

Harada, K., Fujita, N., Murakami, T., *et al.* (1988). An MRI study of CSF flow at the level of C1. *Seminar on Magnetic Resonance Measurements*, p. 86

Ida, M., Hata, Y., Tada, S. and Abe, T. (1988). Cine MR imaging of cerebrospinal fluid flow in syringomyelia. *Radiology*, **169**(P), 198

James, A. E., Jr., DeBlanc, H. J., Jr., DeLand, F. H. and Mathews, E. S. (1972). Refinements in cerebrospinal fluid diversionary shunt evaluation by cisternography. *Am. J. Radiol.*, **115**, 766–773

James, A. E., Jr., DeLand, F. H., Hodges, F. J., III and Wagner, H. N., Jr. (1970). Normal-pressure hydrocephalus: role of cisternography in diagnosis. *J. Am. Med. Assoc.*, **213**, 1615–1622

James, A. E., Flor, W. J., Bush, M., Merz, T. and Rish, B. (1974a). An experimental model for chronic communicating hydrocephalus. *J. Neurosurg.*, **41**, 32–37

James, A. E., Jr., Flor, W. J., Novak, G. R., *et al.* (1977). Experimental hydrocephalus. *Exp. Eye Res.*, **25** (Suppl.), 435–459

James, A. E., Jr., Novak, G. R., Strecker, E.-P., Burns, B. B., Correa-Paz, F. and Flor, W. J. (1980). Experimental studies relating to diagnostic imaging in disorders of cerebrospinal fluid circulation. In Wood, J. H. (Ed.), *Neurobiology of Cerebrospinal Fluid*, Vol. 1. Plenum Press, New York, pp. 381–404

James, A. E., Jr., Strecker, E.-P., Bush, R. M. and Merz, T. (1973). Use of silastic to produce communicating hydrocephalus. *Invest. Radiol.*, **8**, 105–110

James, A. E., Jr., Strecker, E.-P., Sperber, E., Flor, W. J., Merz, T. and Burns, B. (1974b). An alternative pathway of cerebrospinal fluid absorption in communicating hydrocephalus: transependymal movement. *Radiology*, **111**, 143–146

Jolesz, F. A., Patz, S., Hawkes, R. C. and Lopez, I. (1987). Fast imaging of CSF flow/motion patterns using steady-state free precession (SSFP). *Invest. Radiol.*, **22**, 761–771

Kahn, T. and Mueller, E. (1989). Measurement of spinal cerebrospinal fluid flow with the real-time acquisition and evaluation of motion technique. *Radiology*, **173**(P), 404

Kawahara, K., Yoshikawa, A., Maeda, M., *et al.* (1988). Cine MR imaging of cerebrospinal fluid flow. *Radiology*, **169**(P), 200

Kraft, K. A., Fatouros, P. P. and Fei, D. Y. (1990). MR bolus-tracking flow imaging: comparison of several methods. *Seminar on Magnetic Resonance Imaging*, p. 14

Levy, L. M., Bolsten, B. D., McVeigh, E. and Bryan, R. N. (1990). MRI of spinal cord and CSF motion with spatial modulation of magnetization. *Seminar on Magnetic Resonance Imaging*, p. 61

Levy, L. M., Di Chiro, G., McCullough, D. C., Dwyer, A. J., Johnson, D. L. and Yang, S. S. L. (1988a). Fixed spinal cord: diagnosis with MR imaging. *Radiology*, **169**, 773–778

Levy, L. M., Di Chiro, G., Patronas, N. J. and Ogawa, T. (1988b). Cerebrospinal fluid spaces and neuraxis: MR imaging of flow and motion relationships and communications. *Radiology*, **169**(P), 70

Levy, L. M., McVeigh, E. R., Zerhouni, E. A., *et al.* (1988c). Presaturation tagging of neuraxis motion and cerebrospinal fluid circulation. *Radiology*, **169**(P), 385

Martin, A. J., Drake, J. M., Lemaire, C. and Henkelman, R. M. (1989). Cerebrospinal fluid shunts: flow measurements with MR imaging. *Radiology*, **173**, 243–247

Mascalchi, M., Ciraolo, L., Bucciolini, M., Inzitari, D., Arnetoli, G. and Dal Pozzo, G. (1989). Fast multiphase imaging of aqueductal CSF flow in demented patients with hydrocephalus. SMRM, p. 12

Milhorat, T. H., Clark, R. G., Hammock, M. K., *et al.* (1970). Structural, ultrastructural, and permeability changes in the ependyma and surrounding brain favoring equilibration in progressive hydrocephalus. *Arch. Neurol.*, **22**, 397–407

Moran, P. R. (1982). A flow velocity zeugmatographic interlace for NMR imaging in humans. *Magn. Reson. Imag.*, **1**, 197–203

Nadel, L., Braun, I. F., Kraft, K., Laine, F. J. and Jensen, M. E. (1989). Clinical utility of MR phase imaging in neuroradiology. *Radiology*, **173**(P), 383

Partain, C. L., Price, R. R., Patton, J. A., *et al.* (1988a). *Magnetic Resonance Imaging*, 2nd edn, Vols 1 and 2. Saunders, Philadelphia

Partain, C. L., Price, R. R., Patton, J. A., Kulkarni, M. V. and James, A. E., Jr. (Eds) (1988c). *Magnetic Resonance Imaging*. Vol. 1: *Clinical Principles*, 2nd edn. Saunders, Philadelphia

Partain, C. L., Price, R. R., Patton, J. A., Kulkarni, M. V. and James, A. E., Jr. (Eds) (1988c). *Magnetic Resonance Imaging*. Vol. 2: *Physical Principles and Instrumentation*, 2nd edn. Saunders, Philadelphia

Partain, C. L., Scatliff, J. H., Staab, E. V., *et al.* (1978). Quantitative multiregional CSF kinetics using serial, metrizamide enhanced computed tomography. *J. Comput. Assist. Tomogr.*, **2**, 467–470

Partain, C. L. and Staab, E. V. (1979). Brain: imaging of cerebrospinal fluid: computed tomography and nuclear medicine correlation. In Sodee, D. B. (Ed.), *Correlation in Diagnostic Imaging: Nuclear Medicine, Ultrasound and Computed Tomography in Medical Practice*. Appleton-Century-Crofts, New York

Price, D. L., James, A. E., Jr., Sperber, E. and Strecker, E.-P. (1976). Communicating hydrocephalus: cisternographic and neuropathologic studies. *Arch. Neurol.*, **33**, 15–20

Price, R. R., Pickens, D. R., Smith, G., Patton, J. A., Partain, C. L. and James, A. E., Jr. (1987). Blood flow assessment with magnetic resonance imaging. *Proceedings of SPIE— The International Society for Optical Engineering*, **767**, 47–54

Quencer, R. M., Hinks, R. S., Post, M. J. D. and Calabro, G. (1989). Intracranial flow of cerebrospinal fluid: qualitative and quantitative evaluation with cine MR imaging. *Radiology*, **173**(P), 115

Randall, C. P., Collins, A. G., Young, I. R., *et al.* (1983). Nuclear magnetic resonance imaging of posterior fossa tumors. *Am. J. Neuroradiol.*, **4**, 1027–1034

Ridgway, J. P., Turnbull, L. W. and Smith, M. A. (1987). Demonstration of pulsatile cerebrospinal-fluid flow using magnetic resonance phase imaging. *Br. J. Radiol.*, **60**, 423–427

Rubin, J. B. and Enzmann, D. R. (1988). Differential flip-angle imaging of cerebrospinal fluid flow: clinical application in the evaluation of syringomyelia. *Radiology*, **169**(P), 70–71

Sherman, J. L. and Citrin, C. M. (1986). Magnetic resonance demonstration of normal CSF flow. *Am. J. Neuroradiol.*, **7**, 3–6

Shimizu, K., Matsuda, T., Sakurai, T., *et al.* (1986). Visualization of moving fluid: quantitative analysis of blood flow velocity using MR imaging. *Radiology*, **159**, 195–199

Strecker, E.-P., James, A. E., Koningsmark, B. and Merz, T. (1974). Autoradiographic observations in experimental communicating hydrocephalus. *Neurology*, **24**, 192–197

Strecker, E.-P., Novak, G. R. and James, A. E., Jr. (1977). Compartmental analysis of cerebrospinal fluid–blood albumin transfer: consideration of kinetics in normal animals and animals with chronic communicating hydrocephalus. *Europ. Neurol.*, **16**, 203–212

Strecker, E.-P., Scheffel, U., Kelley, J. E. T. and James, A. E., Jr. (1973). Cerebrospinal fluid absorption in communicating hydrocephalus: evaluation of transfer of radioactive albumin from subarachnoid space to plasma. *Neurology*, **23**, 854–864

Thomsen, C., Stahlberg, F., Mogelvang, J., Stubgaard, M. and Nordell, B. (1989). Fourier analysis of cerebrospinal fluid flow in cerebral aqueduct. *Radiology*, **173**(P), 115

Van den Hout, J. H. W., Bakker, C. J. G., Mali, W. P. T. M., *et al.* (1989). Magnetic resonance imaging of the cerebral aqueduct: signal intensity time curves demonstrated by fast acquisition with multiple excitation (FAME). *Invest. Radiol.*, **24**, 855–860

Vessal, K., Sperber, E. E. and James, A. E., Jr. (1974). Chronic communicating hydrocephalus with normal CSF pressures: a cisternographic-pathologic correlation. *Ann. Radiol.*, **17**, 785–793

Waluch, V. and Bradley, W. G. (1984). NMR even echo rephasing in slow laminar flow. *J. Comput. Assist. Tomogr.*, **8**, 594–598

Wesbey, G. E., Moseley, M. E. and Ehman, R. L. (1984). Transitional molecular self-diffusion in magnetic resonance imaging: effects and applications. In James, T. L. and Margulis, A. R. (Eds), *Biomedical Magnetic Resonance*. University of California Press, San Francisco, pp. 63–78

14
Opening of the Blood–Brain Barrier to D-Mannitol Induced by Sensorimotor Cortical Lesions in the Anaesthetized Guinea-pig

Lj. M. Rakić, B. V. Zloković, Jasmina B. Mačkić, M. N. Lipovac, D. M. Mitrović, R. Veskov, Z. Redžić, M. B. Segal and Hugh Davson

INTRODUCTION

We applied the brain vascular perfusion method to determine the kinetics of entry to inert polar molecular D-[^{3}H]-mannitol into the parietal cortex, hippocampus and caudate nucleus in guinea-pigs with sensorimotor (SM) and occipital (Oc) cortical lesions. In the control group of animals, with Oc lesions, only a moderate increase of blood–brain (BBB) permeability to D-mannitol was found after the ipsilateral lesion and no permeability changes occurred after the contralateral lesion, which was expressed by graphically determined unidirectional transfer constant, K_{in}. In a separate series of experiments, in the guinea-pigs with SM cortical lesions, estimated K_{in} values indicated a marked increase of blood-to-brain transport of D-[^{3}H]-mannitol in all brain regions studied, the changes being significantly higher after the contralateral lesion. Amphetamine pretreatment in the guinea-pigs with SM and Oc cortical lesions induced more pronounced BBB permeability changes for D-[^{3}H]-mannitol in all brain areas studied.

MATERIALS AND METHODS

Experiments were carried out on adult guinea-pigs anaesthetized with thiopentone sodium (30 mg/kg i.p.). The sensorimotor cortex was unilaterally removed on either the right or the left side to the depth of white matter (4 mm lateral from midline, 2 mm anterior and 3 mm posterior to bregma) by suction lesion. 'Control' lesions were induced in a similar way in the occipital lobe. Animals were subjected to the BBB permeability studies at different times following the SM cortical lesions (2–32 days) and 2 days after the Oc cortical lesions.

In a separate series of experiments animals with SM lesions were treated with D-amphetamine sulphate (5 mg/kg, daily), and animals with Oc lesions with a single dose of D-amphetamine (5 mg/kg, 90 min before the perfusion). These animals were also subjected to the BBB permeability studies 2–32 days following the lesion.

BBB permeability studies were performed by means of the recently developed technique for vascular perfusion of the ipsilateral forebrain of the guinea-pig *in vivo* (Zloković *et al.*, 1986). The permeability of the BBB to D-mannitol in animals with either ipsilateral or contralateral cortical lesions and in amphetamine-treated animals was expressed by the unidirectional capillary transfer constant K_{in}, as reported previously (Zloković *et al.*, 1986).

RESULTS

Table 14.1 represents values for the unidirectional transfer constant, K_{in}, for D-[^{3}H]-mannitol in different regions of vascularly perfused guinea-pig brain with either ipsilateral or contralateral lesions of the occipital cortex. A moderate increase of the BBB permeability was observed only after the ipsilateral lesion, changes being significantly higher (compared with values for non-treated animals by analysis of variance) in the parietal cortex; K_{in} values estimated in distant brain regions did not differ significantly from those obtained in the corresponding regions of non-treated animals. Acute amphetamine pretreatment greatly increased BBB permeability to D-mannitol in the animals with both ipsilateral and contralateral occipital cortical lesion.

Table 14.2 demonstrates K_{in} values obtained for D-mannitol in the vascularly perfused caudate nucleus after Oc or SM ipsilateral and contralateral cortical lesions. K_{in} values obtained in animals with Oc lesions compared with values for non-treated animals do not exhibit any significant change. In contrast, a significant increase of BBB permeability was found 2 days after SM cortical lesion, the increase being more pronounced after the contralateral lesion. The increase was potentiated by amphetamine after both ipsilateral and contralateral lesion. From 7 to 32 days after the lesion a further increase of BBB permeability was found, changes again being greater after contralateral lesion and potentiated by amphetamine treatment.

Table 14.1 Graphically estimated unidirectional transfer constant K_{in}, for D-[^{3}H]-mannitol in different brain regions in guinea-pigs following ipsilateral or contralateral lesion of the occipital cortex, 2 days after the lesion

Brain region	K_{in} (ml min^{-1} g^{-1} · 10^3)		
	Non-treated animals	Ipsilateral lesion	Contralateral lesion
Hippocampus	0.24 ± 0.02	0.46 ± 0.07 3.30 ± 0.85 (A)	0.37 ± 0.05 1.76 ± 0.58 (A)
Caudate nucleus	0.22 ± 0.02	0.46 ± 0.19 2.31 ± 0.50 (A)	0.23 ± 0.05 2.01 ± 0.43 (A)
Parietal cortex	0.22 ± 0.02	0.60 ± 0.05 3.51 ± 1.33 (A)	0.33 ± 0.08 1.83 ± 0.71 (A)

Control values for non-treated animals taken from Rakić *et al.* (1988a). Effects of amphetamine (A) on K_{in} values have been studied in a separate series of experiments. Values are means $\pm$ SE; number of animals, 3–5.

Table 14.2 Graphically estimated unidirectional transfer constant K_{in}, for D-[^{3}H]-mannitol in the caudate nucleus of the perfused guinea-pig brain following ipsilateral and contralateral cortical (occipital and sensorimotor) lesion

Cortical lesion	Days after lesion	K_{in} (ml min^{-1} g^{-1} · 10^3)	
		Ipsilateral lesion	Contralateral lesion
Occipital	2	0.46 ± 0.19 2.31 ± 0.50 (A)	0.23 ± 0.05 2.01 ± 0.43 (A)
Sensorimotor	2	0.82 ± 0.08 1.30 ± 0.05 (A)	1.22 ± 0.25 2.00 ± 0.41 (A)
	7	1.06 ± 0.02 1.80 ± 0.32 (A)	3.16 ± 0.66 4.36 ± 0.25 (A)
	32	2.56 ± 0.31 3.14 ± 0.10 (A)	2.46 ± 0.12 4.58 ± 1.25 (A)

Effects of amphetamine (A) on K_{in} values have been studied in a separate series of experiments. Values are means $\pm$ SE; number of animals, 3–6.

DISCUSSION

This study has demonstrated an opening of the BBB to inert polar molecule, D-mannitol, in different brain regions after either ipsilateral or contralateral SM cortical lesion (Table 14.2). The BBB permeability to D-mannitol was significantly higher on the contralateral side. The opening of the BBB in distant brain regions after SM lesion most likely demonstrates a lack of trophic influence to maintain the BBB phenomena in other parts of the brain that are normally enriched by sensorimotor cortical input (Rakić *et al.*, 1988b), while the 'neutral' lesion (occipital cortex) did not produce significant influence on the BBB permeability to D-mannitol. Whether this opening represents a compensatory phenomenon remains to be explored further.

It has been shown previously that amphetamine treatment induces reversible breakdown of the BBB, most likely by opening channels that permit a flow of fluid carrying substances at a rate irrespective of their molecular size and lipophilicity (Rakić *et al.*, 1988a). The effect of amphetamine to potentiate the increase of BBB permeability after the cortical lesion could be of importance for recovery at early stages after the lesion.

REFERENCES

Rakić, Lj. Zloković, B. V., Davson, H., Begley, D. J., Segal, M. B., Mitrović, D. M., Mačkić, J. B. and Veskov, R. (1988a). Experimental psychosis and transport of amino acids and peptides across the blood–brain barrier. In Rakić, Lj., Begley, D. J., Davson, H. and Zloković, B. V. (Eds), *Peptide and Amino Acid Transport Mechanisms in the Central Nervous System*. Macmillan Press, London, pp. 169–181

Rakić, Lj., Zloković, B. V., Segal, M. B., Lipovac, M. N., Mitrović, D. M., Veskov, R., Mačkić, J. B. and Davson, H. (1988b). Effects of sensorimotor cortical lesions on the blood–brain barrier permeability to guinea-pigs. *Metab. Brain Dis.*, **4**(1), 9–15

Zloković, B. V., Begley, D. J., Djuricić, B. and Mitrović, D. V. (1986). Measurement of solute transport across the blood–brain barrier in the perfused guinea-pig brain: Method and application of *N*-methyl-α-aminobutyric acid. *J. Neurochem.*, **46**, 1444–1451

15
Uptake of Thiamine by the Isolated Perfused Sheep Choroid Plexus

Dushka M. Mitrović, M. B. Segal, J. E. Preston, H. Davson and
B. V. Zloković

INTRODUCTION

The importance of B group vitamins, especially thiamine, in brain metabolism has been shown by certain neurological disorders resulting from a thiamine deficiency. Thiamine is an essential vitamin in most mammals, and has an important role as a coenzyme in carbohydrate metabolism.

In the present study we have investigated the possibility that $[^{14}C]$-thiamine hydrochloride may pass the blood–cerebrospinal fluid barrier. We used perfused choroid plexus of the sheep (Segal and Pollay, 1977; Dean and Segal, 1985) and the single-circulation paired-tracer dilution technique (Yudilevich and Mann, 1982; Yudilevich, 1986).

MATERIAL AND METHODS

The procedure of the choroid plexus isolation has already been described by Dean and Segal (1985). Briefly, the brain was removed from sheep weighing 25–30 kg anaesthetized by i.v. thiopentone sodium (20 mg/kg). Both internal carotid arteries and the great vein of Galen were cannulated, while all other vessels arising from the circle of Willis were tied off, except the anterior choroidal arteries. The roofs of both lateral ventricles were reflected to expose the perfused choroid plexuses, which were then superfused with artificial cerebrospinal fluid (CSF). The venous drainage from both plexuses was collected from cannula placed in the great vein of Galen.

Thiamine transport was studied by means of the high-resolution paired-tracer dilution method (Yudilevich and Mann, 1982; Yudilevich, 1986). A 100 μl bolus containing the radioactive mixture of 1 μCi D-[^{3}H]-mannitol (extracellular reference) and 0.2 μCi [^{14}C]-thiamine hydrochloride was introduced into either left or right choroid artery. This was followed by collection of 20 drops of successive samples from the great vein of Galen. Standards and one-drop samples were then prepared for liquid scintillation counting.

Cellular uptake at the blood–tissue interface of the choroid plexus of [^{14}C]-thiamine was estimated directly by comparing venous profiles of ^{14}C and ^{3}H radioactivities, using the expression

$$\text{uptake} = [1 - ([^{14}\text{C}]\text{-thiamine}/[^3\text{H}]\text{-mannitol})] \cdot 100$$

The maximal cellular uptake (U_{max}) was calculated from the average of the 'plateau area' of the uptake curve (Figure 15.1). Net cellular uptake (U_{net}) was calculated from the equation:

$$U_{net} = [1 - (\text{total }^{14}\text{C activity recovered/total }^3\text{H activity recovered})] \cdot 100$$

RESULTS

The results of our experiments are given in Table 15.1 and Figure 15.1.

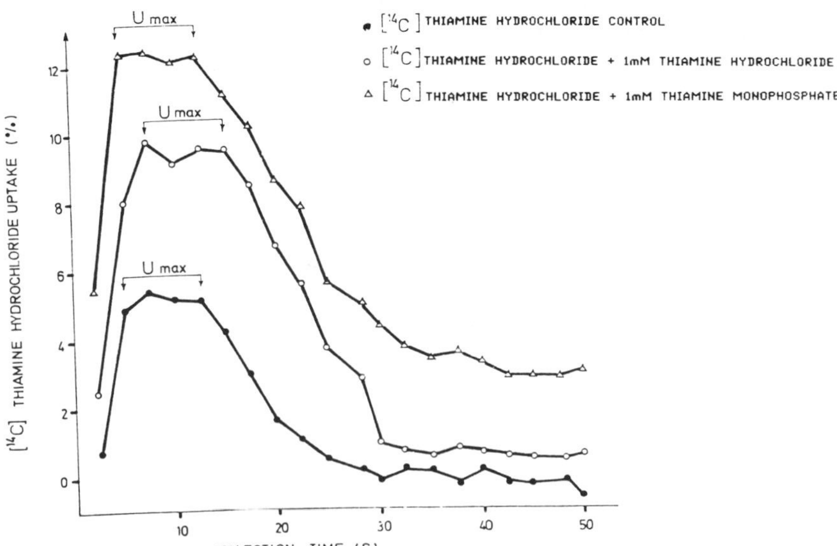

Figure 15.1 Effects of competitors on unidirectional [^{14}C]-thiamine hydrochloride uptake in the isolated perfused choroid plexus of the sheep. Control U_{max} (•) of ^{14}C radioactivity is presented relative to D-[^{3}H]-mannitol, as well as in the presence of unlabelled diamine hydrochloride (1 mM, ○) and thiamine monophosphate (1 mM, △). The uptake values are plotted against the time of collection

Table 15.1 Blood–CSF barrier uptake of thiamine during a single capillary passage through the isolated perfused choroid plexus of the sheep

Tracer	Maximal cellular uptake (U_{max})	% of U_{max} intensification	Net cellular uptake (U_{net})	% of U_{net} intensification	n
[^{14}C]-Thiamine hydrochloride	3.2 ± 3.3		0.80 ± 0.87		9
+1 mM Thiamine hydrochloride	$11.0 \pm 0.5^*$	70.9	0.92 ± 0.70	12.9	7
+1 mM Thiamine monophosphate	$12.2 \pm 1.8^*$	73.8	$2.40 \pm 0.72^*$	66.7	7

Values are mean $\pm$ SEM; n is number of experiments. Probability by Student's t-test: $^* p < 0.05$.

DISCUSSION

Our experiments have demonstrated a very low cellular uptake of thiamine hydrochloride in the single circulation time. This could be due to its low specific activity, such that the injectate concentration was about 9 µM. This, in our experiments the lowest possible concentration, is far above the reported affinity constant ($K_m < 1$ µM) for intestinal (Rose, 1987) and blood–brain barrier transport (Greenwood *et al.*, 1982). In spite of this low uptake value, we could not show a saturable mechanism of thiamine transport from blood to CSF. The entry step into the cell is probably due to diffusion through the membrane, since partitional coefficient olive oil–water for this vitamin is 0.053 (Mitrović *et al.*, 1988). In the presence of unlabelled thiamine hydrochloride (1 mM) and thiamine monophosphate (1 mM) we have shown significantly higher U_{max} of [^{14}C]-thiamine hydrochloride ($11.0 \pm 0.5\%$ and $12.2 \pm 1.8\%$, respectively). We found low control values for U_{net}; this means that just 1% of injected dose of thiamine is retained in the plexus after a single pass. Furthermore, there is an increase of U_{net} in the presence of unlabelled thiamine hydrochloride and thiamine monophosphate. In order to explain the low control values for U_{net}, at least two theoretical explanations can be postulated. First, this method is a single pass of thiamine through the choroid plexus, so this vitamin is retained for a very short time with high dilution. Second, there could be carrier-mediated transport for this vitamin from CSF to blood.

We can conclude that there is a saturable mechanism of thiamine transport from CSF to blood, but the influx into the cell from the blood side is due to the diffusion or by a carrier which is less susceptible to saturation by the substrate.

REFERENCES

Dean, R. and Segal, M. B. (1985). The transport of sugars across the perfused choroid plexus of the sheep. *J. Physiol. (London)*, **362**, 245–260

Greenwood, J., Love, E. R. and Pratt, O. E. (1982). Kinetics of thiamine transport across the blood–brain barrier in the rat. *J. Physiol. (London)*, **327**, 95–103

Mitrović, D. M., Zloković, B. V., Lipovac, M. N. and Davson, H. (1988). Permeability of the blood–brain barrier to (^{14}C)-thiamine in perfused guinea pig brain. *Iugoslav. Physiol. Pharmacol. Acta*, **24**, 273–274

Rose, R. C. (1987). Intestinal absorption of water-soluble vitamins. In Johnson, L. R. (Ed.), *Physiology of the Gastrointestinal Tract*. Raven Press, New York, pp. 1581–1596

Segal, M. B. and Pollay, M. (1977). The secretion of cerebrospinal fluid. In Bito, L. Z., Davson, H. and Fenstermacher, J. D. (Eds), *The Ocular and Cerebrospinal Fluids. Exp. Eye Res.*, **25**, 205–228

Yudilevich, D. L. (1986). Characterization of membrane carriers and receptors at the blood-tissue interface in various organs by single circulation paired tracer dilution. In Yudilevich, D. L. and Mann, G. E. (Eds), *Carrier-mediated Transport of Solutes from Blood to Tissue*. Longman, London, pp. 1–31

Yudilevich, D. L. and Mann, G. E. (1982). Unidirectional uptake of substrates at the blood-tissue interface of secretory epithelia: stomach, salivary gland and pancreas. *Fed. Proc.* **41**, 3045–3054

16
Epileptogenic Activity of Metaphit-induced Audiogenic Seizure in Small Rodents

Milo N. Lipovac, V. Susić, B. V. Zloković, A. E. Jacobson, K. C. Rice, A. N. Popović, S. Popadić and M. E. A. Reith

INTRODUCTION

Phencyclidine (1-(1-phenylcyclohexyl)-piperidine (PCP)) is a drug with anaesthetic and psychotomimetic properties (Hayes and Balster, 1985). It has been shown that metaphit (1-1-(3-isothiocyanatophenyl-cyclohexyl-piperidine), a derivative of PCP, in *in vitro* studies irreversibly binds to the PCP site of PCP/NMDA (*N*-methyl-D-aspartate) receptor, presumably by an acylation process (Rafferty *et al.*, 1985). Recently, in *in vivo* studies, metaphit has been found to induce audiogenic seizure in mice (Debler *et al.*, 1989).

In this study we demonstrated that sound susceptibility induced by metaphit in mice, rats and guinea-pigs has epileptogenic characteristics.

MATERIALS AND METHODS

Adult male Wistar rats, Swiss Webster mice and guinea-pigs were used. These animals did not respond with seizure to six consecutive auditory stimulations given on days before experiments.

Under sodium pentobarbital anaesthesia, for electrocorticographic (EEG) recordings, four gold-plated screws (tip diameter 0.6 mm) were used; two 2 mm posterior to bregma bilaterally and two 1.5–2 mm anterior to lambda, 1–3 mm left and right of midline. Also, two deep, stainless steel wire electrodes were inserted into hippocampus. Recordings from both layers were desirable: slow waves from cortical and theta from hippocampal. Stainless steel wires, bared at the tip, were inserted into dorsal neck muscle for electromyographic (EMG) recording.

Electrode leads were crimped into female Amphenol mini-connector pins, and fixed to the skull with dental acrylic cement. Recordings were taken on a RIZ model EEG machine, at a paper speed of 0.75 cm/s, or faster when necessary. Two animals of the same species were simultaneously recorded (one to be injected with saline and the other with metaphit). After 48 h recovery from surgery, recordings were made for 8 continuous hours to determine the amount of each behavioural state (wakefulness, slow-wave sleep and paradoxical sleep).

Metaphit (50 mg/kg), dissolved in saline, was injected intraperitoneally (i.p.), and control animals received saline. Auditory stimulation (100 dB) was applied 24 h post-injections, until the seizure episode ended.

During the time interval between the injections and sound stimulation continuous recordings were taken in all animals to determine the EEG and EMG signs of metaphit action.

RESULTS AND DISCUSSION

EEG and EMG recordings, from control and metaphit animal groups, taken in the period after injections and before sound stimulation, were analysed and no difference between saline and metaphit-injected animals in duration of waking and sleep phases was found. In contrast, every one of the metaphit-treated animals had electroencephalographic (EEG) changes.

After approximately 3 h from metaphit administration first EEG changes were formed by isolated sharp waves or spikes. Their incidence increased progressively, forming a series of sharp waves. At first, these spike-and-wave complexes were not accompanied by behavioural motor phenomena. After approximately 10 h EEG spiking activity was followed by myoclonic jerks. Isolated myoclonic twitches of forepaws and tremor of whole body were also seen but these motor phenomena never progressed to clonic or clonic–tonic seizures (Figures 16.1, 16.2, 16.3).

The typical seizure pattern displayed by metaphit-treated animals in response to sound stimulation (24 h after metaphit injection) consisted of the EEG and behavioural phenomena. EEG started with desynchronization in response to sound and then changed into very frequent sharp waves, spike-and-wave complexes and polyspikes. These EEG changes corresponded to the latency from the initiation of sound to the onset of motor phenomena starting with wild running followed by clonic and clonic–tonic seizure. Defaecation and micturition were usually seen during that phase. In some animals paroxysmal EEG activity lasted long after the termination of motor phenomena, and at the end of spiking activity profound EEG

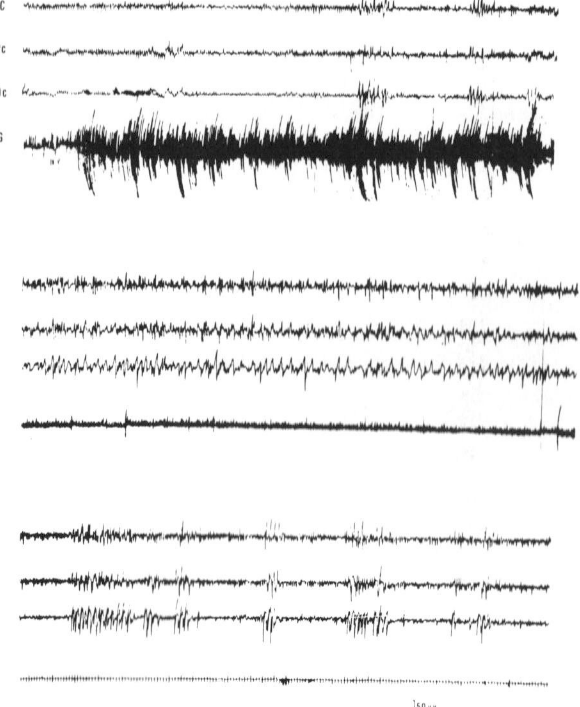

Figure 16.1 Spike and spike-and-wave complexes recorded from a rat 22 h after metaphit (50 mg/kg i.p.) at waking (upper four traces), slow-wave sleep (middle four) and paradoxical sleep (bottom four). RFLc, right–left frontal cortex; RLPc, right–left parietal cortex; RLOc, right–left occipital cortex; EMG, electromyogram of the neck muscle

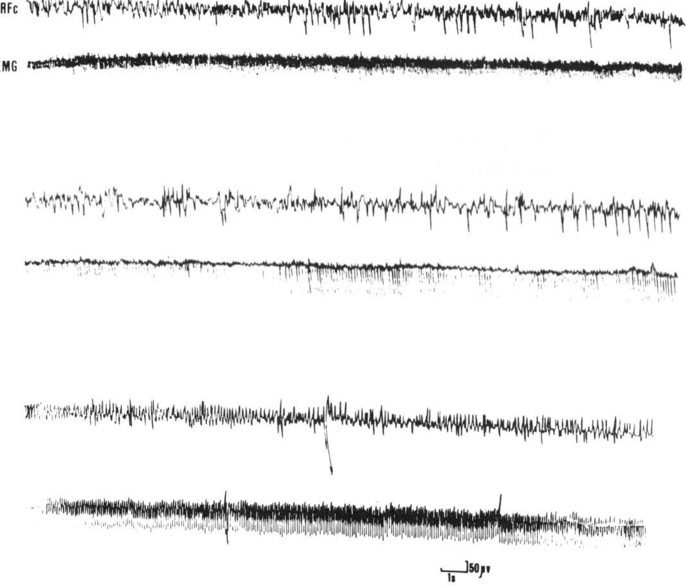

Figure 16.2 Spike and spike-and-wave complexes recorded from a mouse 20 h after metaphit (50 mg/kg i.p.) at waking (upper two traces), slow-wave sleep (middle two) and paradoxical sleep (bottom two). LFRc, left–right frontal cortex; EMG, electromyogram of the neck muscle

depression was seen, followed by full recovery to the preconvulsion pattern (Figures 16.4, 16.5, 16.6).

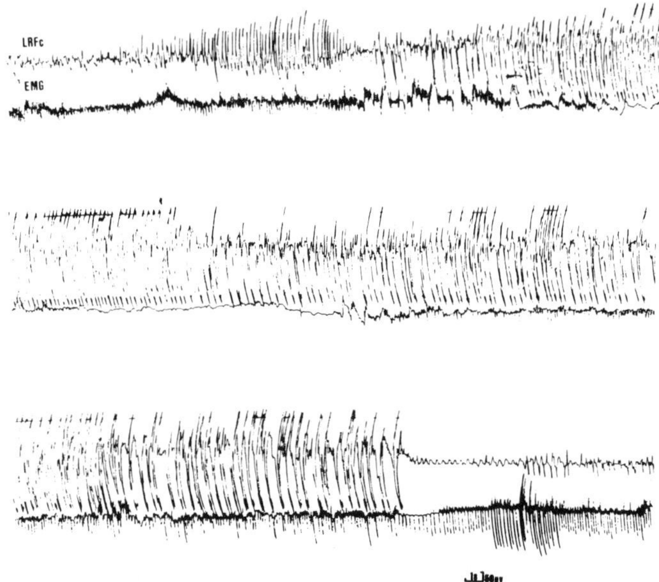

Figure 16.3 Electrocorticographic seizure activity recorded from a guinea-pig 22 h after metaphit (50 mg/kg i.p.) not accompanied by overt motor phenomena. The upper column continues to the lower. LRFc, left–right frontal cortex; EMG, electromyogram of the neck muscle

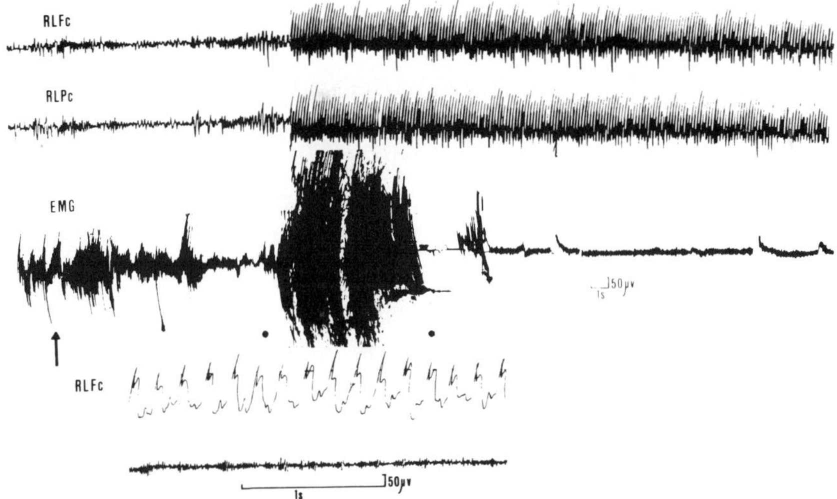

Figure 16.4 Typical record of audiogenic seizure of a rat. Seizure was induced by sound stimulation 24 h after metaphit (50 mg/kg i.p.). Arrow denotes the onset of running, clonic and tonic convulsions. Note the faster paper speed at the bottom of the figure to show configuration of spike wave complexes

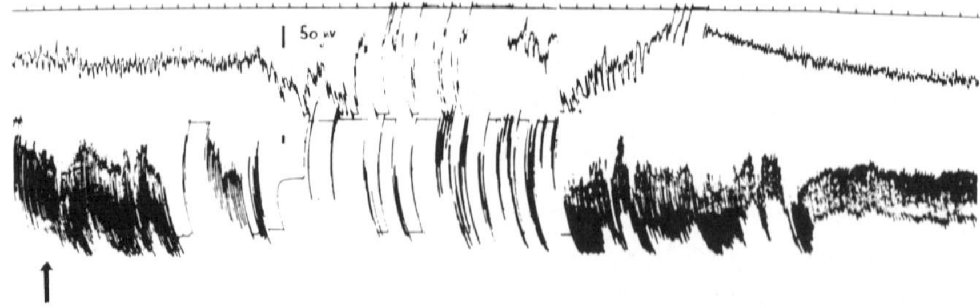

Figure 16.5 Audiogenic seizure in a mouse 24 h after metaphit (50 mg/kg i.p.) administration. Arrow denotes sound stimulus. Upper trace: time in seconds. Middle trace: electrocorticogram. Bottom trace: electromyogram of the neck muscle

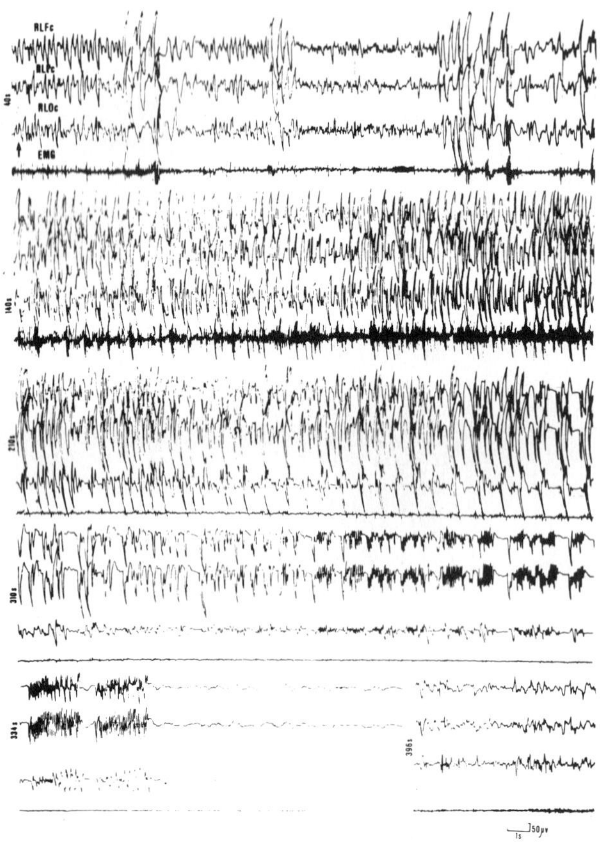

Figure 16.6 Seizure record from a guinea-pig following sound stimulation 24 h after metaphit (50 mg/kg i.p.). Arrow denotes the onset of the sound. Numbers denote time in seconds after sound stimulation. Note depression and then recovery of EEG at 396 s

It is possible that the metaphit-induced epileptogenic activity is the result of an enhanced probability that the NMDA/PCP ion channel is open. Metaphit may irreversibly bind to a PCP site allosterically linked to the NMDA receptor, which increases the affinity of that site for the naturally present ligand, probably glutamate. Also, metaphit could induce an upregulation of NMDA receptors, and such increase in receptor number would make the animals sensitive to seizure long after the metaphit injection.

SUMMARY

Adult male rats, mice and guinea-pigs were prepared for chronic recording of EEG and EMG. After recovery, animals were exposed to the sound (100 dB) for recording control EEG and EMG, and observation of normal behavioural phenomena.

After that, one group received metaphit (50 mg/kg) and the other saline. Animals were monitored for 24 h and exposed to the sound. The behaviour, EEG and EMG results were compared.

All metaphit-treated animals showed evidence of spike-and-wave and paroxysmal activity which became more rapid after sound stimulation. These EEG changes were associated with myoclonic jerks, followed by wild running, which rapidly developed into generalized clonic–tonic seizure after the sound.

REFERENCES

Debler, E. A., Lipovac, M. N., Lajtha, A., Zloković, B. V., Jacobson, A. E., Rice, K. C. and Reith, M. E. A. (1989). Metaphit, an isothiocyanate analog of PCP, induces audiogenic seizures in mice. *Europ. J. Pharmacol.*, **165**, 155–159

Hayes, B. A. and Balster, R. L. (1985). Anticonvulsant properties of phencyclidine-like drugs in mice. *Europ. J. Pharmacol.*, **117**, 121–123

Rafferty, M. F., Mattson, M., Jacobson, A. E. and Rice, K. C. (1985). A specific acylating agent for the (3H)phencyclidine receptors in rat brain. *FEBS Lett.*, **181**, 318–321

17
Anticonvulsant Effects of Phencyclidine and PCP-like Drugs on Audiogenic Seizures Induced by Metaphit in Mice

Milo N. Lipovac, E. A. Debler, A. Lajtha, B. V. Zloković,
A. E. Jacobson, K. C. Rice, A. N. Popović, S. Popadić and
M. E. A. Reith

INTRODUCTION

Phencyclidine (PCP), at low doses (0.3–3 mg/kg), selectively delayed the onset or prevented the occurrence of pentylenetetrazol (PTZ)-induced tonic, but not clonic seizures. However, at a dose of 10 mg/kg it acted as a convulsant agent (Hayes and Balster, 1985). PCP-like drugs, when evaluated with respect to potential anti/pro-convulsant actions, demonstrated only anticonvulsant properties. Modulation of the N-methyl-D-aspartate (NMDA) receptor site by PCP and PCP-like compounds as a mechanism of anticonvulsant activity has been suggested by Leander et $al.$ (1988), in which a strong correlation was demonstrated between the anticonvulsant activity of PCP/PCP-like compounds and the minimal effective dose which blocks NMDA receptor-mediated NMDA-induced lethality.

In metaphit an acylating function has been added to the PCP molecule by placing an isothiocyanate group in the meta position on the aromatic ring. In behavioural studies, PCP-induced stereotypy and ataxia in rats was antagonized by metaphit administration 24 h prior to PCP (Contreras et $al.$, 1985), an effect accompanied by a reduction in the number of PCP binding sites in the brain (Contreras et $al.$, 1986). Some of the behavioural effects of metaphit appear to be species- and/or time-dependent. Metaphit did not antagonize catalepsy in pigeons induced by administration of PCP 24 h later. However, 4 h after injection metaphit itself induced PCP-like catalepsy. In Rhesus monkeys, metaphit induced both ataxia and, 2 h after injection, spontaneous clonic convulsions (Koek et $al.$, 1986).

MATERIALS AND METHODS

Adult BALB/cBy mice weighing 20–25 g were used. The animals were kept on a 12 h light/dark cycle (07:00 to 19:00 light), with food and water available ad libitum. Auditory stimulation was applied using an electric bell generating 100 dB. The sound stimulus was presented until the seizure episode ended or for a period of time not longer than 60 s. All compounds studied were injected intraperitoneally (i.p.) at a volume of 0.1 ml. The drugs used were: metaphit (methane-sulphonate, synthesized according to the procedure of Wang *et al.*, 1986), MK-801 (5-methyl-10,11-dihydro-5*H*-dibenzo(a,d)cyclohepten-5,10-imine maleate), and phencyclidine (PCP hydrochloride).

RESULTS AND DISCUSSION

Metaphit administered i.p. at a dose of 30 mg/kg produced the occurrence of audiogenic clonic to clonic to clonic/tonic (grades 4–7, according to seizure scale, Figure 17.1) seizures in 80% of the mice exposed to a 100 dB audio stimulus 24 h later. If metaphit was administered at doses of 80 and 120 mg/kg, all of the animals evaluated exhibited seizure behaviour of grades 4–7 (Figure 17.2).

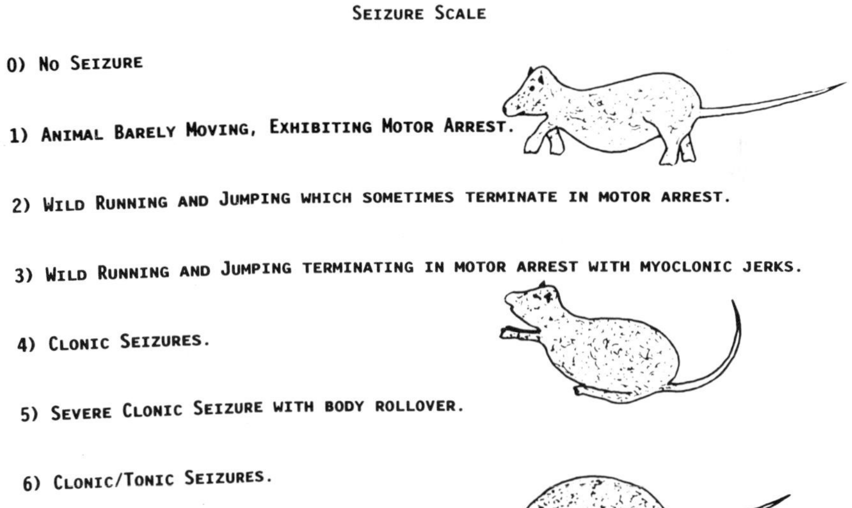

Figure 17.1 Dose–response curve of metaphit-induced seizures. Metaphit was administered i.p. at doses of 5, 10, 20, 30, 40, 60, 80, 120 mg/kg. Twenty-four hours after metaphit administration the mice were exposed to the 100 dB sound. (Number of animals studied for each experimental group was 12–20)

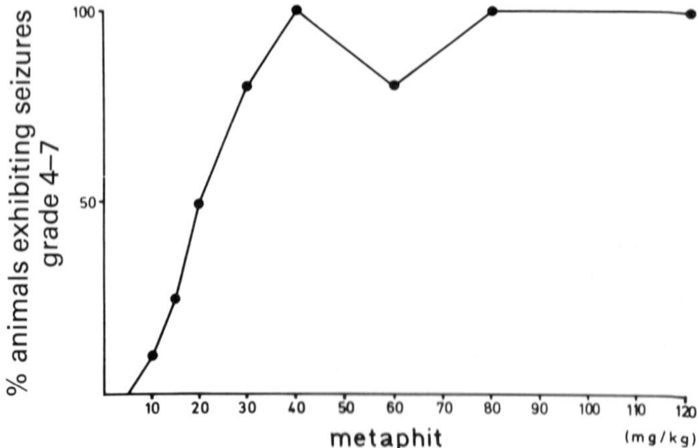

Figure 17.2 Time course of metaphit-induced audiogenic seizures. Animals were exposed to 100 dB sound after metaphit administration (i.p., 120 mg/kg). Mice tested at 1, 2, 3, 4, 5 and 24 h after metaphit administration represent separate groups ($n = 7$–20). These individual groups were evaluated again at 48 h

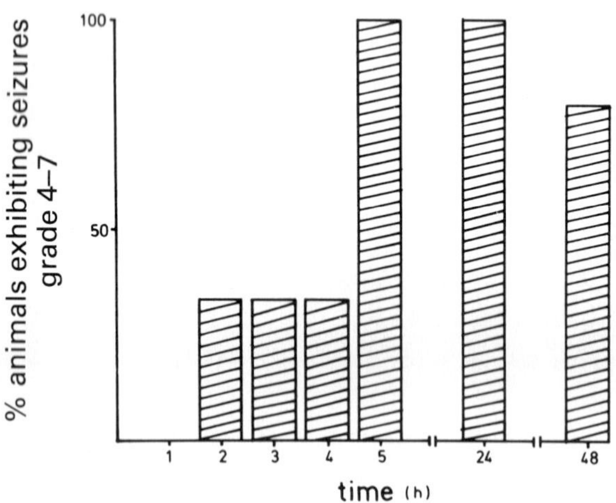

Figure 17.3 Time course of seizures induced by a 100 dB audiogenic stimulus in rats which had received 30 mg/kg metaphit

Time course studies demonstrate that metaphit-induced audiogenic seizures developed with time. One hour after metaphit administration no animal exhibited clonic seizure behaviour upon exposure to the 100 dB stimulus. Two to four hours after metaphit injection only 33% of the animals exposed to the sound stimulus exhibited clonic seizure behaviour. After 5 h audiogenic (grade 4–7) seizures occurred in all the animals

Table 17.1 Metaphit-induced audiogenic seizures 24 h after administration. Seizures were induced by a 100 dB sound 24 h after an i.p. injection of metaphit. The animal groups were as follows: (1) control, saline injected i.p.; (2) metaphit administered i.p. at a dose of 80 mg/kg; (3) PCP injected (i.p., 5 mg/kg); (4) MK-801 (0.5 mg/kg) injected i.p. 30 min prior to metaphit injection (i.p., 80 mg/kg); (5) PCP injected (i.p., 5 and 7 mg/kg) 24 h after metaphit (i.p., 80 mg/kg) and 30 min before exposure to the 100 dB; (6) MK-801 (0.5 mg/kg) administered 24 h after metaphit (i.p., 80 mg/kg) and 30 min prior to audio stimulation; (7) either PCP (5 mg/kg) or MK-801 (0.5 mg/kg) administered i.p. 24 h prior to audio stimulation

Treatment	N_c/N_t	Percentage of animals exhibiting grade 4–7 seizures
Saline	0/15	0
PCP (5 mg/kg)	0/8	0
MK-801 (0.5 mg/kg)	0/8	0
Metaphit (80 mg/kg)	20/20	100
Mk-801 (0.5 mg/kg) +metaphit (80 mg/kg)	10/10	100
PCP (5 mg/kg) +metaphit (80 mg/kg)	10/10	100
Metaphit (80 mg/kg) +MK-801 (0.5 mg/kg)	0/10	0
Metaphit (80 mg/kg) +PCP (5 mg/kg)	2/10	20
Metaphit (80 mg/kg) +PCP (7 mg/kg)	0/5	0

N_c, number of animals convulsing; N_t number of animals treated.

evaluated (Figure 17.3). After 48 h the seizure behaviour begins to diminish (Debler *et al.*, 1989).

The occurrence of metaphit-induced seizures was blocked by either PCP nor MK-801 administered 24 h after metaphit injection and 30 min prior to the sound. However, neither PCP, nor MK-801 proved effective in the inhibition of the seizures when administered 30 min prior to the metaphit (Table 17.1).

It is possible that the metaphit-induced seizures are the results of an enhanced probability that the ion channel is open. Bound PCP appears to act as a channel blocker inhibiting (Ca^{2+}, Na^+ and K^+) movement either by its physical presence or by modifying the gating. The anticonvulsant activity of PCP and PCP-like drugs can be related to this blocking characteristic.

SUMMARY

Metaphit administration makes mice susceptible to audiogenic seizures and sensitivity to audio stimulus increases during the first 5 h after metaphit administration. After 48 h the alterations induced by metaphit begin to abate and the mice become less sensitive to the audio stimulation. These audiogenic seizures can be prevented by the administration of either PCP or MK-801 prior to the audio stimulation. Administration of PCP or MK-801 30 min prior to metaphit does not protect mice 24 h later from the effects of metaphit. Phencyclidine or MK-801 itself does not make mice susceptible to audiogenic seizures 24 h after its administration.

REFERENCES

Contreras, P. C., Johnson, S., Freedman, R., Hoffer, B., Olsen, K., Rafferty, M. F., Lessor, R. A., Rice, K. C., Jacobson, A. E. and O'Donohue, T. L. (1986). Metaphit, an acylating ligand for phencyclidine receptors: characterization of *in vivo* actions in the rat. *J. Pharmacol. Exp. Ther.*, **238**, 1101–1104

Contreras, P. C., Rafferty, R. A., Lessor, R. A., Rice, K. C., Jacobson, A. E. and O'Donohue, T. L. (1985). A specific alkylating ligand for phencyclidine (PCP) receptors antagonizes PCP behavioral effects. *Europ. J. Pharmacol.*, **111**, 405–408

Debler, E. A., Lipovac, M. N., Lajtha, A., Zloković, B. V., Jacobson, A. E., Rice, K. C. and Reith, M. E. A. (1989). Metaphit, an isothiocyanate analog of PCP, induces audiogenic seizures in mice. *Europ. J. Pharmacol.*, **165**, 155–159

Hayes, B. A. and Balster, R. L. (1985). Anticonvulsant properties of phencyclidine like drugs in mice. *Europ. J. Pharmacol.*, **117**, 121–123

Koek, W., Woods, J. H., Jacobson, A. E., Rice, K. C. and Lessor, R. A. (1986). Metaphit, a proposed phencyclidine receptor acylator: phencyclidine-like behavior effects and evidence of absence of antagonist activity in pigeons and in Rhesus monkeys. *J. Pharmacol. Exp. Ther.*, **237**, 386–390

Leander, J. D., Rathbun, R. C. and Zimmerman, D. M. (1988). Anticonvulsant effect of phencyclidine-like drugs: relation to *N*-methyl-D-aspartic acid antagonism. *Brain Res.*, **454**, 368–371

Wang, Y., Palmer, M., Freedman, R., Hoffer, B., Mattson, M. V., Lessor, R. A., Rice, K. C. and Jacobson, A. E. (1986). Antagonism of phencyclidine action by metaphit in rat cerebellar purkinje neurons: an electrophysiological study. *Proc. Natl Acad. Sci. USA*, **83**, 2724–2727

18
Cotransport of Sodium, Potassium and Chloride in the Isolated Choroid Plexus

C. Johanson, J. Parmelee, D. Bairamian, S. Sweeney and M. Epstein

INTRODUCTION

The linkage between cation and Cl transport in the form of Na–Cl or Na–K–Cl cotransporters is a widespread phenomenon in animal cells (Haas and McManus, 1983; Palfrey and Rao, 1983; Tas *et al.*, 1987). Scant attention, however, has been paid to the mammalian choroid plexus (CP) in regard to the function, or even the existence, of cation–Cl cotransport. Pharmacologic investigations of the CP–CSF system of intact animals have furnished evidence for cation–Cl cotransport in some experiments (McCarthy and Reed, 1974; Melby *et al.*, 1982), but not in others (Vogh and Langham, 1981; Miller *et al.*, 1986). We have recently used an isolated preparation to analyse ion transport in CP (Parmelee and Johanson, 1989). Such an *in vitro* analysis permits fine control of artificial CSF concentrations of ions and drugs over a wide range (Bairamian *et al.*, 1990; Johanson *et al.*, 1990). This approach has enabled us to obtain solid evidence for Na–K–Cl cotransport in rat CP.

METHODS

Lateral ventricle CP tissues were removed from adult Sprague–Dawley rats (200–350 g), preincubated with or without drug for 20 min, and then transferred to artificial CSF (aCSF) medium containing ^{22}Na, ^{86}Rb(K) or ^{36}Cl with [^{3}H]-mannitol (extracellular correction). Complete baseline curves were established for the cellular uptake of Na, Cl and K at 37 °C (Bairamian *et al.*, 1990; Johanson *et al.*, 1990). The half-maximal uptake was used to assess the effects of drugs or ion substitution on the rate of transport. The osmolality, ion composition and glucose content of the aCSF was as previously described (Smith *et al.*, 1981). The aCSF was gassed with 95% O_2/5% CO_2 and held close to a pH of 7.35.

RESULTS

The transport of Na and Cl in *in vitro* CP cells was so fast that by 12 s the uptake was about one-half that at steady state (Figure 18.1). The loop diuretic agents, bumetanide and furosemide, were both tested for their ability to inhibit NaCl transport (Figure 18.2). Bumetanide (0.01 mM) reduced the uptake of Na and Cl by 28% and 45%, respectively. Furosemide was less potent in its effects. Thus, at 0.1 mM, it suppressed the uptake of Na and Cl by 25% and 33%.

Potassium uptake was also significantly reduced by 0.01 mM bumetanide and 0.1 mM furosemide (results not shown). In other experiments, the ouabain-insensitive K uptake was stimulated by first preincubating the CP in very low Na and K aCSF, then incubating at higher concentrations of Na and K. The K uptake stimulated in this way was substantially blocked by 0.01 mM bumetanide. Moreover, the preincubation of CP in Na- or Cl-free medium, followed by incubation in normal aCSF, stimulated the ouabain-insensitive uptake of K by 133%. This stimulated K transport required *both* Na and Cl in aCSF (Bairamian *et al.*, 1990).

DISCUSSION

Na–Cl or Na–K–Cl cotransport is involved in the transport of salt and water across many secretory epithelia. A common feature of these cation–Cl cotransporters is their inhibition by furosemide and bumetanide. Furosemide at 0.1 mM decreased NaCl uptake in CP by about 30%. At a tenfold lower concentration, bumetanide reduced CP uptake of Na and Cl to an even greater degree. Lesser inhibitory potency of furosemide (vs. bumetanide) to inhibit Na–Cl cotransport has been found in other epithelia such as trachea (Widdicombe *et al.*, 1983) and rectal gland (Palfrey *et al.*, 1984). Overall, the ability of both furosemide and bumetanide to suppress Na as well as Cl transport in rat CP, together with their relative inhibitory potencies, constitute pharmacological evidence for cation–Cl cotransport.

At relatively high concentrations, certain diuretic drugs lose their selectivity for interfering with particular transporters. At concentrations used in this study, however, these loop diuretics would be expected to inhibit primarily Na–K–Cl cotransport but not to interfere significantly with antiporters such as Na–H and Cl–HCO_3 (Palfrey *et al.*, 1984). We previously demonstrated clear-cut *additive* inhibition (>90%) of CP up-

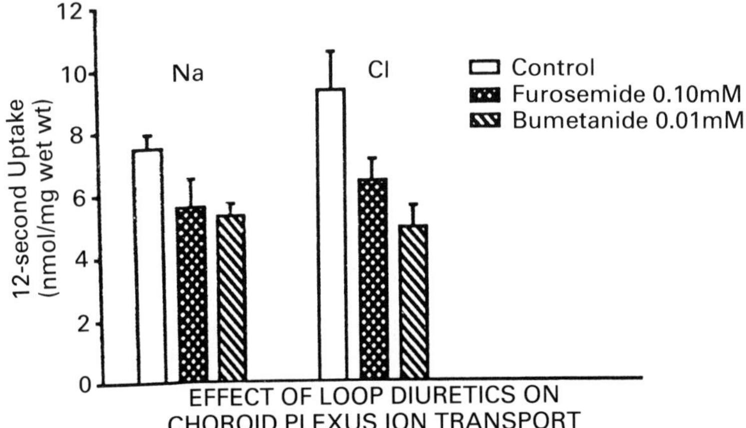

Figure 18.1 Effects of furosemide and bumetanide on the *in vitro* CP uptake of Na and Cl. Each bar is the mean ± SE for 7 or more CP tissues. All drug treatments are significantly different from corresponding controls; $p < 0.05$ by multiple range test

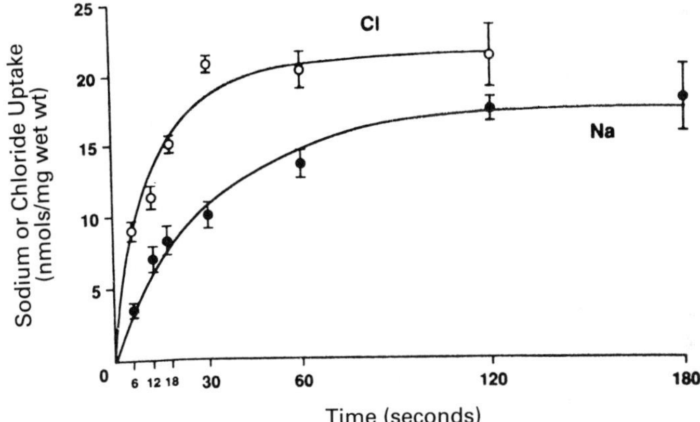

Figure 18.2 Time course of uptake of Na and Cl by *in vitro* lateral ventricle CP. Uptake is corrected for extracellular label. Each symbol is the mean for 7 or more CPs, except for the steady-state values, for which $n = 4$ or more tissues

take of Cl, caused by the inclusion of 0.1 mM DIDS with bumetanide in the aCSF incubation medium (Johanson *et al.*, 1990). DIDS is a disulphonic stilbene inhibitor of Cl–HCO$_3$ exchange, but it is ineffective against Na–K–Cl cotransport (Palfrey *et al.*, 1984). Collectively, such findings indicate that bumetanide inhibits a transporter in CP that is separate from the DIDS-sensitive one.

Potassium is essential for NaCl cotransport in many epithelial tissues (Haas, 1989). We found that the loop diuretic-sensitive, ouabain-insensitive K transport had an absolute dependence on both Na and Cl in the incubation medium. Therefore, the CP is another tissue in which K is an obligatorily cotransported species. Because some cells exhibit K-independent NaCl cotransport (perhaps an alternative mode of Na–K–Cl under some physiological conditions), we cannot rule out the presence or potential in CP for NaCl cotransport without K.

SUMMARY

By using drug inhibitors with or without CSF ion substitution, we have found that a portion of the Na, K and Cl uptake by mammalian CP is sensitive to both bumetanide and furosemide. These findings point to the existence of Na–K–Cl cotransport in adult rat CP. Earlier studies evaluating effects of loop agents on CSF dynamics inferred cation–Cl cotransport in CP (Reed, 1969; Melby *et al.*, 1982). The present, more direct approach with isolated CP has produced findings corroborating Na–K–Cl cotransport. Currently we are using cisternal microdialysis to evaluate the effects of loop diuretics on the presumed component of CSF formation that stems from ion cotransport across CP membranes.

REFERENCES

Bairamian, D., Johanson, C. E., Parmelee, J. T. and Epstein, M. H. (1990). Potassium cotransport with sodium and chloride in the choroid plexus. Submitted to *J. Neurochem.*

Haas, M. (1989). Properties and diversity of (Na–K–Cl) cotransporters. *Ann. Rev. Physiol.*, **51**, 443–457

Haas, M. and McManus, T. J. (1983). Bumetanide inhibits (Na + K + 2 Cl) cotransport at a chloride site. *Am. J. Physiol.*, **245**, C235–C240

Johanson, C. E., Sweeney, S. M., Parmelee, J. T. and Epstein, M. H. (1990). Cotransport of sodium and chloride by the adult mammalian choroid plexus. *Am. J. Physiol.*, **258**, C211–C216

McCarthy, K. O. and Reed. D. J. (1974). The effect of acetazolamide and furosemide on cerebrospinal fluid production and choroid plexus carbonic anhydrase activity. *J. Pharmacol. Exp. Ther.*, **189**, 194–201

Melby, J. M., Miner, L. C. and Reed, D. J. (1982). Effect of acetazolamide and furosemide on the production of cerebrospinal fluid from the cat choroid plexus. *Can. J. Physiol. Pharmacol.*, **60**, 405–409

Miller, T. B., Wilkinson, H. A., Rosenfeld, S. A. and Furuta, T. (1986). Intracranial hypertension and cerebrospinal fluid production in dogs: Effects of furosemide. *Exp. Neurol.*, **94**, 66–80

Palfrey, H. C. and Rao, M. C. (1983). Na/K/Cl co-transport and its regulation. *J. Exp. Biol.*, **106**, 43–54

Palfrey, H. C., Silva, P. and Epstein, F. H. (1984). Sensitivity of cAMP stimulated salt secretion in the shark rectal gland to 'loop' diuretics. *Am. J. Physiol.*, **246**, C242–C246

Parmelee, J. T. and Johanson, C. E. (1989). Development of potassium transport capability by choroid plexus of infant rats. *Am. J. Physiol.*, **256**, R786–R791

Reed, D. J. (1969). The effect of furosemide on cerebrospinal fluid flow in rabbits. *Arch. Int. Pharmacodyn. Ther.*, **178**, 324–330

Saito, Y. and Wright, E. M. (1987). Regulation of intracellular chloride in bullfrog choroid plexus. *Brain Res.*, **417**, 267–272

Smith, Q. R., Woodbury, D. M. and Johanson, C. E. (1981). Uptake of Cl-36 and Na-22 by the choroid plexus–cerebrospinal fluid system: Evidence for active choroid transport by the choroidal epithelium. *J. Neurochem.*, **37**, 107–116

Tas, P. W., Massa, P. T., Kress, H. G. and Koschel, K. (1987). Characterization of an $Na^+/K^+/Cl^-$ co-transport in primary cultures of rat astrocytes. *Biochim. Biophys. Acta*, **903**, 411–416

Vogh, B. P. and Langham, M. R., Jr. (1981). The effect of furosemide and bumetanide on cerebrospinal fluid formation. *Brain Res.*, **221**, 171–183

Widdicombe, J. H., Nathanson, I. T. and Highland, E. (1983). Effects of 'loop' diuretics on ion transport by dog tracheal epithelium. *Am. J. Physiol.*, **245**, C388–C396

19
Effect of Congenital Hydrocephalus on Cortical Structure in the H-Tx Rat

R. M. Bucknall, N. G. Harris and H. C. Jones

The H-Tx rat develops severe congenital hydrocephalus with an incidence of approximately 70%. The condition develops in late gestation and is noticeable outwardly after birth by doming of the skull. It progresses rapidly, with onset of symptoms appearing after weaning and death at 6–7 weeks of age. The purpose of this study was to determine the effects of hydrocephalus on cortical thickness and the developing microvasculature and cell density (neurons + glia) in the cerebral cortex.

Fixed tissue was taken from postnatal rats at 10 days (mid-stage hydrocephalus) and 30 days (late-stage hydrocephalus), and five regions of the cerebral cortex (Figure 19.1) were prepared for light microscopy in one of two ways: whole brain embedding in wax followed by coronal sections at 6 μm or small pieces excised, embedded in epoxy resin and sectioned at 1 μm. A Seescan image analyser connected to a Leitz Dialux microscope via a video camera was used to analyse both types of section. Capillary number was measured on the 1 μm sections and capillary numerical density (N_a, number of capillaries per unit area: Weibel, 1979) was calculated for the different brain regions. Cortical thickness and cell density were measured on the wax sections. The cell densities were corrected for section thickness, using the correction factor of Abercrombie (1946).

Capillary N_a was significantly reduced in the hydrocephalics at 30 days after birth ($p < 0.01$) but not at 10 days after birth (Figure 19.1A). Cortical thickness was reduced by hydrocephalus in all regions at both ages by 10–80%, especially in the posterior regions (auditory and visual), with the worst of the thinning already present by 10 days after birth (Figure 19.1B). Cell density was significantly reduced in the hydrocephalics by 23–40% at 30 days after birth but not at 10 days after birth (Figure 19.1C).

Thus, in spite of severe thinning of the cerebral cortex at 10 days after

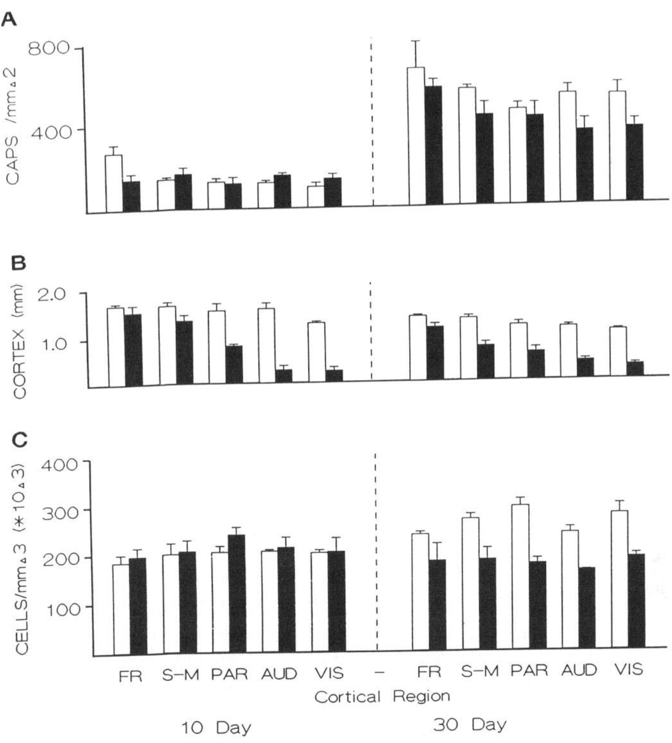

Figure 19.1 (A) Capillary numerical density; (B) cortical thickness; (C) cell density. Control rats (open bars) and hydrocephalic rats (shaded bars) at 10 and 30 days after birth. Values are means ± SEM. FR, frontal cortex; S-M, sensorimotor cortex; PAR, parietal cortex; AUD, auditory cortex; VIS, visual cortex

birth, the reduction in capillary numerical density and cell density does not occur until 30 days after birth, when the hydrocephalus is advanced.

REFERENCES

Abercrombie, M. (1946). Estimations of nuclear populations from microtome sections. *Anat. Rec.*, **94**, 239–247

Weibel, E. R. (1979). *Stereological Methods*, Vol. 1. Academic Press, London

20
Role of the Blood–Brain Barrier in Immunopathogenesis of Experimentally Induced Autoimmune Demyelination

D. S. Skundrić, B. V. Zloković, M. B. Segal, Lj. Rakić and H. Davson

Experimental allergic encephalomyelitis (EAE) has been studied as experimental model for human primary demyelinating diseases such as multiple sclerosis. There are a number of experimental data indicating altered function of the blood–brain barrier (BBB) during the acute and chronic-relapsing forms of EAE (Bradbury, 1989).

In the present study in guinea-pigs with an acute EAE induced by injection of myelin basic protein (MBP), transport of blood-borne immunoglobulin G (IgG) from blood into the brain across the BBB was studied by the vascular brain perfusion method (Zloković, 1986) followed by immunohistochemical analysis of the brain tissue.

EAE was induced in Hartly albino guinea-pigs, weighing 300–350 g, using the following procedure (Colover *et al.*, 1989): (a) on day 1, guinea-pigs were injected s.c. with water-in-oil emulsion containing 15 μg of muramyl dipeptide, 200 μg of ovalbumin in Freund incomplete adjuvant; (b) on day 28, they were injected i.p. with 10 mg of ovalbumin in saline under Piritone protection; and (c) on day 35, survivors were injected with 50 μg of homologous MBP emulsified in 0.1 ml of Freund complete adjuvant. Guinea-pigs were observed daily in order to obtain a clinical score during a period of up to 45 days after MBP injection.

Vascular brain perfusion was performed in anaesthetized guinea-pig forebrain *in situ*. After surgical exposure of the neck blood vessels, the right carotid artery was cannulated and connected to the extracorporeal perfusion circuit; at the start of the vascular perfusion, both jugular veins were cut and the contralateral carotid artery ligated. Perfusion pressure was kept within the physiological range. Perfusion was started by the artificial blood for about 5 min, composed of an artificial plasma in which

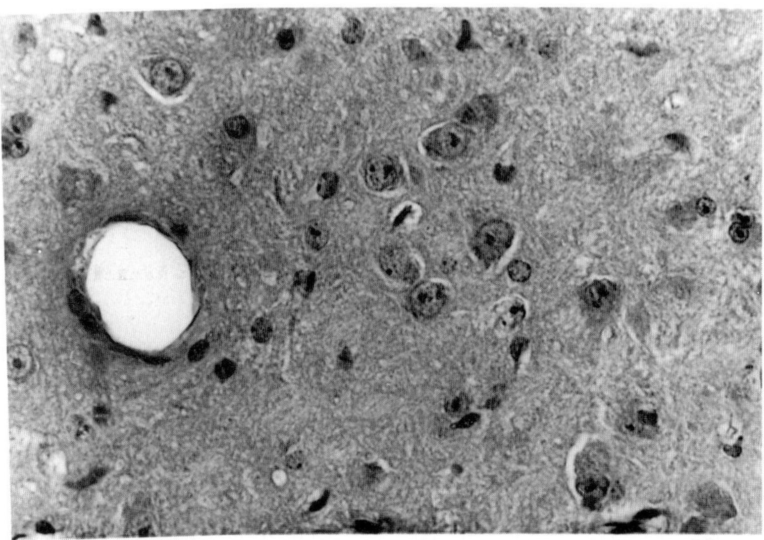

Figure 20.1 Positive perivascular peroxidase reaction after 10 minutes' perfusion with homologous IgG (4 mg/ml) in the region of the hippocampus in the EAE guinea-pig

sheep red blood cells were suspended with haematocrit 20% (Zloković, 1986). Perfusion was continued with homologous IgG in a concentration corresponding to the physiological level of IgG in guinea-pig serum for 10 min. Perfusion was terminated by fixative (2.5% glutaraldehyde) which was perfused for 15 min. After that the guinea-pig was decapitated, and the perfused brain hemisphere was removed and suspended in the same fixative for 4 h and then prepared for immunohistochemical analysis. The described procedure was performed in: EAE guinea-pigs (when manifesting clinical signs of EAE) and normal, control guinea-pigs. In normal guinea-pigs, two control groups were examined. One control was: normal guinea-pig perfused with artificial blood and then with fixative. Another control was: normal guinea-pig perfused with artificial blood, then with IgG and after that with fixative (in the same way as EAE guinea-pigs).

The avidin-biotin-peroxidase method was performed on paraffin-embedded sections of the perfused brain tissue in order to determine the presence of the homologous IgG in the guinea-pig brain. After deparaffination and dehydration, endogenous peroxidase was blocked by methanol and slides of the brain tissue were incubated with 10% guinea-pig serum diluted in phosphate-buffered saline (PBS). The procedure involved incubations with anti-guinea-pig IgG, biotinylated anti-anti IgG, avidin-peroxidase and finally specific substrate for peroxidase reaction, diamino benzidine. Slides were contrastained in haematoxylin, dewaxed, cleared and mounted for light microscopy examination.

In a sample of 30 randomly chosen guinea-pigs, immunized with homologous MBP, about 60% developed clinical signs of EAE (from weight loss to paraparesis and paraplegia of hind limb). In this group of animals, vascular brain perfusion and immunohistochemical analysis were performed.

An intensive immunoperoxidase reaction, after 10 min perfusion with homologous IgG, with locations at endothelial cells of the forebrain capillaries, surrounding perivascular spaces, ependyma and choroid plexi, was observed in EAE guinea-pigs. Normal guinea-pigs not perfused with IgG did not show any positive peroxidase reaction, while normal guinea-pigs perfused with IgG showed weak positive peroxidase staining related to endothelial cells but not in the surrounding perivascular spaces (Figure 20.1).

The results of the present study support our previously reported results (Zloković *et al.*, 1989; Skundrić *et al.*, 1990) indicating that BBB may play an important role in immunoglobulin homoeostasis within the central nervous system during experimentally induced autoimmune demyelination.

REFERENCES

Bradbury, M. W. (1989). Transport across the blood-brain barrier. In Neuwelt, E. A. (Ed.), *Implication of the Blood-Brain Barrier and its Manipulation*, Vol. 1, Plenum, New York, pp. 119–137

Colover, J., Skundrić, D. S. and Zloković, B. (1989). Chain of events leading to demyelination. In Gonsette, R. E. and Delmotte, P. (Eds), *Recent Advances in Multiple Sclerosis Therapy*. Elsevier Science Publishers, B.V., pp. 305–308

Daniel, P. M., Lam, D. K. C. and Pratt, O. E. (1981). Changes in the effectiveness of the blood–brain and blood–spinal cord barrier in experimental allergic encephalomyelitis. *J. Neurol. Sci.*, **52**, 211–219

Skundrić, D. S., Zloković, B. V., Colover, J., Dumonde, D. and Rakić, Lj. (1990). Transport of homologous immunoglobulin G across the blood–brain barrier during EAE in the guinea-pig. *Period. Biol.*, **92**, 1

Zloković, B. V., Begley, D. J., Djurić, B. and Mitrović, D. M. (1986). Measurements of solute transport across the blood-brain barrier in the perfused guinea pig brain: method and application to N-methyl-α-amino-isobutyric acid, *J. Neurochem.*, **46**, 1444–1459

Zloković, B. V., Skundrić, D. S., Segal, M. B., Colover, J., Jankov, R. M., Pejnović, N., Lacković, V., Mackić, J., Lipovac, M., Davson, H., Kasp, E., Dumonde, D. and Rakić, Lj. (1989). Blood–brain barrier permeability changes during acute allergic encephalomyelitis induced in the guinea-pig. *Metab. Brain Dis.*, **4**(1), 33–40

Index